Nutritional Neurosciences

Series Editor
Mohamed Essa, Sultan Qaboos University, Qaboos, Oman

This book series aims to publish volumes focusing on both basic and clinical research in the field of nutritional neuroscience with a focus on delineating the effect of nutrition on brain function and behavior. The books will examine the role of different nutrients, food agents and supplements (both macro and micro) on brain health, neurodevelopment, neurochemistry, and behaviour. The books will examine the influence of diet, including phytochemicals, antioxidants, dietary supplements, food additives, and other nutrients on the physiology and metabolism of neurons, neurotransmitters and their receptors, cognition, behavior, and hormonal regulations.

The books will also cover the influence of nutrients and dietary supplements on the management of neurological disorders. It details the mechanism of action of phytonutrients on signaling pathways linked with protein folding, aggregation, and neuroinflammation. The books published in the series will be useful for neuroscientists, nutritionists, neurologists, psychiatrists, and those interested in preventive medicine.

Hamdan Hamdan
Editor

Exploring the Effects of Diet on the Development and Prognosis of Multiple Sclerosis (MS)

Editor
Hamdan Hamdan
Department of Biological Sciences
College of Medicine and Health Sciences, Khalifa University
Abu Dhabi, United Arab Emirates

ISSN 2730-6712 ISSN 2730-6720 (electronic)
Nutritional Neurosciences
ISBN 978-981-97-4672-9 ISBN 978-981-97-4673-6 (eBook)
https://doi.org/10.1007/978-981-97-4673-6

© The Editor(s) (if applicable) and The Author(s), under exclusive license to Springer Nature Singapore Pte Ltd. 2024

This work is subject to copyright. All rights are solely and exclusively licensed by the Publisher, whether the whole or part of the material is concerned, specifically the rights of translation, reprinting, reuse of illustrations, recitation, broadcasting, reproduction on microfilms or in any other physical way, and transmission or information storage and retrieval, electronic adaptation, computer software, or by similar or dissimilar methodology now known or hereafter developed.

The use of general descriptive names, registered names, trademarks, service marks, etc. in this publication does not imply, even in the absence of a specific statement, that such names are exempt from the relevant protective laws and regulations and therefore free for general use.

The publisher, the authors and the editors are safe to assume that the advice and information in this book are believed to be true and accurate at the date of publication. Neither the publisher nor the authors or the editors give a warranty, expressed or implied, with respect to the material contained herein or for any errors or omissions that may have been made. The publisher remains neutral with regard to jurisdictional claims in published maps and institutional affiliations.

This Springer imprint is published by the registered company Springer Nature Singapore Pte Ltd.
The registered company address is: 152 Beach Road, #21-01/04 Gateway East, Singapore 189721, Singapore

If disposing of this product, please recycle the paper.

Preface

This book offers an insightful exploration into the relationship between diet and multiple sclerosis (MS), aiming to address a crucial question: Can dietary interventions serve as effective preventative and therapeutic measures for patients with MS? Delving into this question, this book examines various dietary components and regimens, shedding light on their potential impacts on the progression, relapse rate, and development of MS. It offers readers valuable insights into how dietary choices can influence the management of this condition.

Backed up by evidence gathered from review and clinical trial papers, this book discusses the role of vitamins such as A, B, and D, as well as dietary supplements like caffeine, carnitine, and lipoic acid in benefiting patients with MS. Particular attention is given to the significance of vitamin D in lowering the risk of developing MS and its immunomodulatory effects on the inflammatory processes associated with the disease.

In parallel, this book also addresses the detrimental effects of diets such as the Western or high salt diet (HSD) on MS prognosis, emphasizing how these dietary regimens can harm the gut microbiome and exacerbate inflammatory responses, ultimately promoting demyelination of the central nervous system (CNS). This book then explores alternative dietary approaches that confer a protective effect to the gut microbiome and the CNS, including whole grain, fasting, Mediterranean, and ketogenic diets.

In order to represent MS holistically, this book delves into the epidemiology, pathogenesis, and epigenetics of MS. It offers a thorough examination of the underlying pathophysiology of the disease, driven by the activation of inflammatory markers and the impairment of the immune system, with a particular focus on the role of CD4 T cells and their Th1 and Th17 subtypes. Finally, the current and emerging therapeutic regimens for the treatment of MS are discussed.

This comprehensive resource is an essential read for patients with MS seeking to understand the potential impacts of diet on their overall health, as well as healthcare professionals and researchers interested in exploring dietary interventions for MS management.

Abu Dhabi, United Arab Emirates   Hamdan Hamdan

Acknowledgements

H.H was supported by a faculty startup grant (FSU-2022-002-8474000395) from Khalifa University.

Contents

1. **Introduction to Multiple Sclerosis** 1
   Maitha M. Alhajeri, Rayyah R. Alkhanjari, Sara Aljoudi,
   Nadia Rabeh, Zakia Dimassi, and Hamdan Hamdan

2. **Life Chapters: Navigating Multiple Sclerosis Across Pregnancy,
   Breastfeeding, Epidemiology, and Beyond** 17
   Salsabil Zubedi, Hana Al-Ali, Nadia Rabeh, Sara Aljoudi,
   Zakia Dimassi, and Hamdan Hamdan

3. **The Role of Gut Microbiota in the Pathophysiology of Multiple
   Sclerosis** 45
   Hana Al-Ali, Salsabil Zubedi, Sara Aljoudi, Nadia Rabeh,
   Zakia Dimassi, and Hamdan Hamdan

4. **Western Diet Impact on Multiple Sclerosis** 57
   Hana Al-Ali, Salsabil Zubedi, Nadia Rabeh, Sara Aljoudi,
   Zakia Dimassi, and Hamdan Hamdan

5. **High Salt Diet Impact on MS** 67
   Salsabil Zubedi, Hana Al-Ali, Sara Aljoudi, Nadia Rabeh,
   Zakia Dimassi, and Hamdan Hamdan

6. **From Pasture to Plate: Investigating the Role of Bovine Sources
   in Multiple Sclerosis** 77
   Nadia Rabeh, Sara Aljoudi, Zakia Dimassi, Haya Jasem Al-Ali,
   Khalood Mohamed Alhosani, and Hamdan Hamdan

7. **Role of Vitamins in Multiple Sclerosis** 95
   Haia M. R. Abdulsamad, Amna Baig, Sara Aljoudi, Nadia Rabeh,
   Zakia Dimassi, and Hamdan Hamdan

8	**The Potential Preventive and Therapeutic Role of Vitamin D in MS** .... Rayyah R. Alkhanjari, Maitha M. Alhajeri, Nadia Rabeh, Sara Aljoudi, Zakia Dimassi, and Hamdan Hamdan	107
9	**Role of Dietary Supplements in Multiple Sclerosis** .... Haia M. R. Abdulsamad, Amna Baig, Sara Aljoudi, Nadia Rabeh, Zakia Dimassi, and Hamdan Hamdan	125
10	**Plant-Based Extracts and Antioxidants: Implications on Multiple Sclerosis** .... Azhar Abdukadir, Rawdah Elbahrawi, Nadia Rabeh, Sara Aljoudi, Zakia Dimassi, and Hamdan Hamdan	139
11	**Dietary Regimens: Whole Grains and Multiple Sclerosis** .... Haia M. R. Abdulsamad, Amna Baig, Sara Aljoudi, Nadia Rabeh, Zakia Dimassi, and Hamdan Hamdan	165
12	**Diet and Its Potential Impact on the Prognosis of Multiple Sclerosis: Fasting Diets** .... Amna Baig, Haia M. R. Abdulsamad, Nadia Rabeh, Sara Aljoudi, Zakia Dimassi, and Hamdan Hamdan	175
13	**Diet and Its Potential Impact on the Prognosis of Multiple Sclerosis: Mediterranean Diet** .... Amna Baig, Haia M. R. Abdulsamad, Sara Aljoudi, Nadia Rabeh, Zakia Dimassi, and Hamdan Hamdan	185
14	**Ketogenic Diet: Implications on Multiple Sclerosis** .... Rawdah Elbahrawi, Azhar Abdukadir, Nadia Rabeh, Sara Aljoudi, Zakia Dimassi, and Hamdan Hamdan	195
15	**Epigenetics: Implication on Multiple Sclerosis** .... Rawdah Elbahrawi, Sara Aljoudi, Nadia Rabeh, Zakia Dimassi, Khalood Mohamed Alhosani, and Hamdan Hamdan	207
16	**Therapeutic Strategies and Ongoing Research** .... Azhar Abdukadir, Nadia Rabeh, Sara Aljoudi, Zakia Dimassi, Khalood Mohamed Alhosani, and Hamdan Hamdan	219

Editor and Contributors

About the Editor

Hamdan Hamdan an Assistant Professor at Khalifa University and Visiting Professor at Baylor College of Medicine, specializes in neurodegenerative diseases and autism spectrum disorders. His laboratory aims to understand the mechanisms by which the axon initial segment (AIS) and the nodes of Ranvier are assembled and maintained in their normal health and during injury or their role in neurodegenerative diseases. Any therapeutic intervention of the nervous system must involve proper maintenance and assembly of the node of Ranvier and the AIS. He is part of a pioneering team from Khalifa University, the first in the MENA region to reach the semifinals of the Longitude Prize on Dementia. In addition to his research, he is an expert in designing different serotypes of viruses, leveraging viral vectors as vehicles for delivering therapeutic genes into cells; he utilizes gene editing CRISPR/Cas9 techniques, designs transgenic animal models, and focuses on advancing our understanding and treatment of neurological disorders. His proficiency in this area and his expertise in gene editing using CRISPR/Cas9 position him at the forefront of innovative approaches for studying and potentially treating neurological disorders.

Contributors

Azhar Abdukadir Department of Biological Sciences, College of Medicine and Health Sciences, Khalifa University, Abu Dhabi, United Arab Emirates

Haia M. R. Abdulsamad Department of Biological Sciences, College of Medicine and Health Sciences, Khalifa University, Abu Dhabi, United Arab Emirates

Hana Al-Ali Department of Biological Sciences, College of Medicine and Health Sciences, Khalifa University, Abu Dhabi, United Arab Emirates

Haya Jasem Al-Ali Department of Biological Sciences, College of Medicine and Health Sciences, Khalifa University, Abu Dhabi, United Arab Emirates

Maitha M. Alhajeri Department of Biological Sciences, College of Medicine and Health Sciences, Khalifa University, Abu Dhabi, United Arab Emirates

Khalood Mohamed Alhosani Department of Biological Sciences, College of Medicine and Health Sciences, Khalifa University, Abu Dhabi, United Arab Emirates

Sara Aljoudi Department of Biological Sciences, College of Medicine and Health Sciences, Khalifa University, Abu Dhabi, United Arab Emirates

Rayyah R. Alkhanjari Department of Biological Sciences, College of Medicine and Health Sciences, Khalifa University, Abu Dhabi, United Arab Emirates

Amna Baig Department of Biological Sciences, College of Medicine and Health Sciences, Khalifa University, Abu Dhabi, United Arab Emirates

Zakia Dimassi Department of Medical Sciences, College of Medicine and Health Sciences, Khalifa University, Abu Dhabi, United Arab Emirates

Rawdah Elbahrawi Department of Biological Sciences, College of Medicine and Health Sciences, Khalifa University, Abu Dhabi, United Arab Emirates

Hamdan Hamdan Department of Biological Sciences, College of Medicine and Health Sciences, Khalifa University, Abu Dhabi, United Arab Emirates

Nadia Rabeh Department of Biological Sciences, College of Medicine and Health Sciences, Khalifa University, Abu Dhabi, United Arab Emirates

Salsabil Zubedi Department of Biological Sciences, College of Medicine and Health Sciences, Khalifa University, Abu Dhabi, United Arab Emirates

Chapter 1
Introduction to Multiple Sclerosis

Maitha M. Alhajeri, Rayyah R. Alkhanjari, Sara Aljoudi, Nadia Rabeh, Zakia Dimassi, and Hamdan Hamdan

Abstract Multiple sclerosis (MS) is a chronic demyelinating central nervous system disease caused by acquired autoimmune inflammation, demyelination, and axonal degeneration. It is primarily a disease of young adults. Various genetic and environmental factors have been identified as possible contributors to the disease pathogenesis, diet being a key factor. We hypothesize that diet can serve as a preventive and therapeutic target for patients with MS. This book is dedicated to collect the available evidence that establishes a relationship between diet and MS, thus exploring dietary interventions as a potential avenue for the prevention and treatment of the disease that is accessible across various cultural and socioeconomic backgrounds.

Keywords Multiple sclerosis (MS) · McDonald's Criteria · Immune-mediated Demyelination · Oligodendrocytes · Disease-modifying therapies (DMTs) · Diet

Abbreviations

APC	Antigen-presenting cells
BBB	Blood-brain barrier
CNS	Central nervous system

M. M. Alhajeri · R. R. Alkhanjari · S. Aljoudi · N. Rabeh · H. Hamdan (✉)
Department of Biological Sciences, College of Medicine and Health Sciences, Khalifa University, Abu Dhabi, United Arab Emirates
e-mail: hamdan.hamdan@ku.ac.ae

Z. Dimassi
Department of Medical Sciences, College of Medicine and Health Sciences, Khalifa University, Abu Dhabi, United Arab Emirates

© The Author(s), under exclusive license to Springer Nature Singapore Pte Ltd. 2024
H. Hamdan (ed.), *Exploring the Effects of Diet on the Development and Prognosis of Multiple Sclerosis (MS)*, Nutritional Neurosciences,
https://doi.org/10.1007/978-981-97-4673-6_1

CSF	Cerebrospinal fluid
DMTs	Disease-modifying therapies
EAE	Experimental allergic encephalomyelitis
GWAS	Genome-wide association studies
HLA	Human leukocyte antigen
HSD	High salt diet
IFN-γ	Interferon-gamma
IL-17	Interleukin-17
IL-22	Interleukin-22
IL-23	Interleukin-23
IL-6	Interleukin-6
MHC class I	Major histocompatibility complex class I
MPO	Myeloperoxidases
MRI	Magnetic resonance imaging
MS	Multiple Sclerosis
NHS	Nurses' Health Study
NMSS	National Multiple Sclerosis Society
NO	Nitric oxide
PP	Primary Progressive
PPMS	Primary Progressive MS
RCTs	Randomized control trials
ROS	Reactive oxygen species
RRMS	Relapsing-remitting MS
S1P	Sphingosine 1 phosphate
SPMS	Secondary-Progressive MS
Th1	T helper 1
Th17	T helper 17
TNF-α	Tumor necrosis factor-alpha
TYK2	Tyrosine kinase 2
UV	Ultraviolet
VDR	Vitamin D receptor
WBCs	White blood cells

Learning Objectives
- Define MS and its underlying pathology involving demyelination and axonal degeneration.
- Classify MS into its main subtypes: Relapsing-Remitting, Secondary Progressive, Primary Progressive, and Progressive Relapsing.
- Explain the diagnostic criteria for MS, focusing on the Revised McDonald's Criteria.

- Explore the inflammatory cascade and cellular interactions leading to demyelination and neurodegeneration.
- Evaluate environmental factors like vitamin D deficiency, obesity, and smoking in the development and progression of MS.

1.1 What Is Multiple Sclerosis, and How Does It Present?

Multiple Sclerosis (MS) is an acquired, immune-mediated disease characterized by the demyelination of the central nervous system (CNS). It commonly begins in adulthood, peaking between 20 and 40, with a female:male of 3:1 or greater female predominance (Ford 2020). It is known for its broad spectrum of clinical presentation and severity, radiological appearance, involved gene loci, and response to therapy (Lucchinetti et al. 2000). Among the classical symptoms seen in most patients are sensory deficits, including paresthesia (e.g., tingling, prickling, or "pins and needles" sensation) and hypesthesia (e.g., reduced sensation or numbness). Patients may develop unpleasant sensory experiences like *Lhermitte's symptom,* an electric shock sensation radiating to the lower extremities, usually triggered by neck flexion. MS can also manifest as motor symptoms resulting from upper motor neuron lesions, including limb weakness accompanied by increased spasticity, hyperreflexia, and positive Babinski signs (Doshi and Chataway 2016; Kasper et al. 2018).

MS can also affect the optic nerve, which is embryologically derived from the prosencephalon and myelinated by the oligodendrocytes, resulting in optic neuritis, a subacute, unilateral, painful vision loss, accompanied by periorbital pain aggravated by eye movement and relative afferent pupillary defect. Internuclear ophthalmoplegia is another ocular sign usually encountered in MS, whereby the immune-mediated demyelination damages the medial longitudinal fasciculus, impairing ipsilateral adduction and causing nystagmus of the contralateral abducting eye (Kasper et al. 2018; Nij Bijvank et al. 2019; Petzold et al. 2014). Interestingly, some patients' neurological symptoms can result from or worsened by an increase in the core body temperature, such as in cases of fever, strenuous physical activity, and hot showers. This phenomenon, known as *Uhthoff's phenomenon*, predominantly manifests as temporary painful vision loss, albeit it can occur with any symptom of MS (Kasper et al. 2018; Opara et al. 2016).

1.2 Classification of MS

Based on the natural history of the disease, MS can be divided into several subtypes. Here we will discuss the main four (Fig. 1.1) according to the classification of the US National Multiple Sclerosis Society (NMSS) Advisory Committee (Polman et al. 2011). In relapsing-remitting MS (RRMS), patients experience discrete, gradual-onset attacks evolving over several days and interspaced with periods of

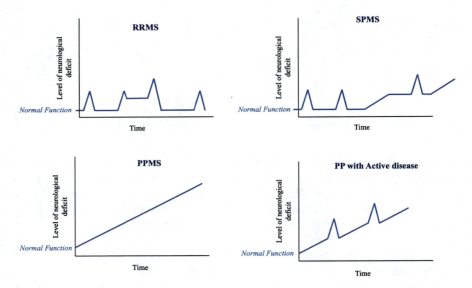

Fig. 1.1 Classification of MS. Each graph represents a disease subtype based on its clinical course. The peaked portion of the graph represents an acute attack. *RRMS* - Relapsing-Remitting MS, *SPMS* - Secondary Progressive MS, *PPMS* - Primary Progressive MS, and *PP with Active disease* - Primary Progressive with Active disease. Adapted from "Clinical presentation and diagnosis of multiple sclerosis," by H. Ford 2020, Clin Med (Lond), 20(4), 380–383. Copyright 2020 by the Royal College of Physicians

neurological stability. The NMSS specifies that symptoms in the RRMS subtype do not develop suddenly and last at least 24 h without signs of fever or infectious process. In secondary progressive MS (SPMS), some patients experience incomplete recovery between the attacks while concomitantly developing gradual worsening of the symptoms after a relapsing-remitting course. SPMS is often diagnosed retrospectively based on the patient's history. Almost 10% of cases present with the primary progressive MS (PPMS) subtype, which is characterized by a gradual, steady decline in neurological function rather than presenting with acute episodic attacks. The fourth subtype of MS is progressive relapsing (PRMS), where the patient demonstrates a steady progressive neurological disability with apparent acute attacks. However, the MS committee recommends categorizing patients with PRMS as Primary Progressive (PP) patients with active disease (Ford 2020; Kasper et al. 2018; Lublin et al. 2014).

1.2.1 MS Diagnosis

Until now, a definitive diagnosis of MS cannot be made based on the results of a single test. A diagnosis is facilitated by the amalgamation of clinical signs and symptoms with various supporting investigations, namely, magnetic resonance imaging (MRI), cerebrospinal fluid (CSF) analysis, and evoked potentials (Doshi and Chataway 2016; Ford 2020). The hallmark feature that the clinical, paraclinical, or both types of investigations must demonstrate to diagnose MS is the spatiotemporal dissemination of the pathologic lesions (Ford 2020; Polman et al. 2011). Based on the most recent Revised McDonald's Criteria, there are various constellations of signs and symptoms that aid in the diagnosis of MS. One approach to reach a diagnosis is to confirm the occurrence of two or more attacks, each lasting for more than 24 h and separated by at least 1 month, and an associated pathology in two or more anatomically unrelated areas in the CNS white matter. Another approach is only a clinical dissemination in time and radiological evidence of dissemination in space that are characteristic for MS. An example of radiological evidence that would support a diagnosis of MS would need to show at least one T2 lesion in at least two out of four CNS areas: periventricular, juxtacortical, infratentorial, or spinal cord. Those are only some of the possibilities that have been identified in patients with a likely diagnosis of MS (refer to Table 1.1 for more information) (Ford 2020; Kasper et al. 2018; Polman et al. 2011).

1.3 Is MS an Autoimmune Disease?

As mentioned previously, MS is an immune-mediated disease driven by a pathological immune response directed against the myelin found in the CNS. Whether it can be classified as an autoimmune disease remains inconclusive. Currently available evidence is not compelling enough to confirm MS as an autoimmune condition. In fact, MS fails to meet several characteristics of the autoimmunity criteria (Lemus et al. 2018; Wootla et al. 2012). Although various studies succeeded in identifying antibodies directed against different CNS antigens such as myelin protein, carbohydrates, and lipids, evidence confirming the presence of specific MS autoantigens is still lacking. Even with autoantibodies recognized, such as those against myelin oligodendrocyte glycoprotein, myelin basic protein, alu repeats, alpha-B-crystalline, and myelin-associated glycoprotein, there is no consistency or consensus among the published studies (Kasper et al. 2018; Lemus et al. 2018). In addition, attempts to induce MS lesions in animal models using the identified antibodies or T cells yielded contrasting findings. For the purpose of lab research in MS, we can generate similar demyelinating disease in animal models using Theiler's murine encephalomyelitis virus, coronavirus, and Semliki Forest virus or through utilizing multiple candidate antigens together (Lemus et al. 2018).

1.4 Pathophysiology of MS

A series of pathophysiological events, including localized immune mononuclear cells infiltration, microglia activation, demyelination, axonal damage, and gliosis, are identified in the pathogenesis of MS (Ciccarelli et al. 2014; Kasper et al. 2018).

Table 1.1 2017 McDonald criteria for the diagnosis of MS in patients with an attack at onset. Adapted from "Clinical presentation and diagnosis of multiple sclerosis," by H. Ford 2020, Clin Med (Lond), 20(4), 380–383. Copyright 2020 by the Royal College of Physicians

Number of attacks at clinical presentation	Number of lesions with objective clinical evidence	Additional data needed for diagnosis of multiple sclerosis
≥2	≥2	None [a]
≥2	1 (as well as clear-cut historical evidence of a previous attack involving a lesion in a distinct anatomical location)	None [a]
≥2	1	Dissemination in space demonstrated by an additional clinical attack implicating a different CNS site
		Or by MRI
1	≥2	Dissemination in time demonstrated by an additional clinical attack
		Or by MRI
		Or demonstration of CSF-specific oligoclonal bands
1	1	Dissemination in space demonstrated by an additional clinical attack implicating a different CNS site
		Or by MRI
		And dissemination in time demonstrated by an additional clinical attack
		Or by MRI
		Or demonstration of CSF-specific oligoclonal bands

[a] = no additional tests are required to demonstrate dissemination in space and time. However, unless MRI is not possible, brain MRI should be obtained in all patients in whom the diagnosis of multiple sclerosis is being considered. In addition, spinal cord MRI or CSF examination should be considered in patients with insufficient clinical and MRI evidence supporting multiple sclerosis, with a presentation other than a typical clinically isolated syndrome, or with atypical features. If imaging or other tests (e.g. CSF) are undertaken and are negative, caution needs to be taken before making a diagnosis of multiple sclerosis, and alternative diagnoses should be considered
CNS central nervous system, *CSF* cerebrospinal fluid, *MRI* magnetic resonance imaging

Among the major immune cells involved are $CD4^+$ (helper) T cells, which have been shown to recognize myelin basic protein based on the experimental allergic encephalomyelitis (EAE) animal model (Choi et al. 2016). Acute inflammation precipitating endothelial-lymphocyte interactions and disrupting the blood-brain barrier (BBB) facilitate the lymphocytic, localized invasion of the CNS seen in MS, as seen in Fig. 1.2. Endothelial-lymphocyte interactions are enabled through the upregulation of endothelial ligands, such as P-selectin and VCAM-1, that bind to

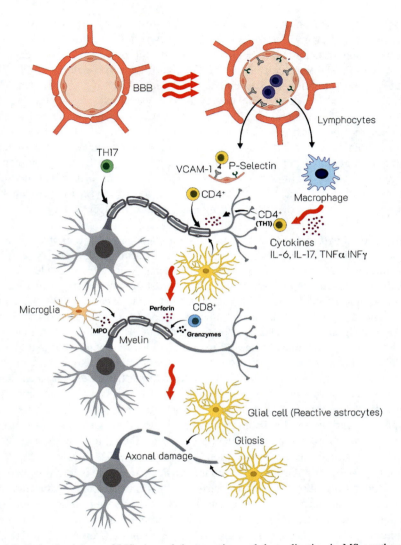

Fig. 1.2 Pathophysiology of MS. Axonal degeneration and demyelination in MS are the combined result of inflammatory reactions from the innate and adaptive immune system. The release of various cytokines, granzymes, perforins, and MPO are the key constituents that drive this process.*A disruption in BBB integrity facilitates the inflammatory process seen in MS. Initially, white blood cells (WBCs) migrate and adhere to cell adhesion molecules, such as VCAM-1 and P-selectin, to reach the CNS and start the sequence of inflammation. Once in the CNS, macrophages, CD4+ T cells, and Th17 release their inflammatory mediators and cytokines. CD8+ T cells also play a role in demyelination by releasing perforins and granzymes, while microglia contribute to this inflammatory cascade by releasing MPO. The consequence of these reactions is myelin sheath deterioration. Scarring of tissue, or gliosis, ensues and is mediated by reactive astrocytes in the CNS

integrins expressed on the surface of lymphocytes. Once lymphocytes cross the BBB, different helper T cells groups, namely type 1 (Th1) and 17 (Th17), their cytokines and inflammatory mediators, such as interleukin-6 (IL-6), IL-17 and IL-22, tumor necrosis factor-alpha (TNF-α), and interferon-gamma (IFN-γ) contribute to demyelination—a pathological hallmark of MS (Ciccarelli et al. 2014; Kasper et al. 2018). B cell lymphocytes also act as follicle-like aggregates in the meninges of SPMS patients, leading to a younger age of onset and contributing to inflammatory mediators associated with profound cortical pathology. In addition to the lymphocytes role in MS, microglial cells contribute to the disease as their activation is triggered by pervasive inflammation and injury in the brain parenchyma. Consequently, they produce myeloperoxidases (MPO) that further fuel the inflammatory response and exacerbate tissue damage. Interestingly, current research suggests the applicability of MPO as a biomarker and therapeutic target (Ciccarelli et al. 2014).

MS pathology encompasses not only myelin damage but also axonal and neuronal degeneration, contributing significantly to the neurological disability seen in the clinical course of the disease. A variety of mechanisms have been identified that explain the neuronal injury seen in MS. By interacting with the major histocompatibility complex class I (MHC class I), CD8+ cytotoxic T cells target demyelinated axons and damage them through the action of perforin and granzymes. Perforin mediates cellular toxicity and apoptosis by creating membranous pores that enable the delivery of granzymes. Studies of neuronal motor function found that MS mice models deficient in perforin demonstrated larger-diameter axons with improved neuronal function (Lemus et al. 2018). Other factors that mediate axonal damage and neuronal loss include reactive oxygen species (ROS) and nitric oxide (NO) released by the activated microglia, infiltrated macrophages, and lymphocytes. Despite the various mechanisms of neuronal injury, damage is mainly caused by mitochondrial dysfunction and reduction in ATP synthesis, leading to oligodendrocytes apoptosis and axon degeneration (Kasper et al. 2018; Lemus et al. 2018). Secondary to this cellular loss and damage, a gliotic response is initiated, manifesting as the proliferation of reactive astrocytes that are detected within and at the border of the inflammatory lesions and in normal-appearing white matter. These reactive astrocytes are hypothesized to be the main contributors to the chronic symptoms of MS as they can interfere with remyelination (Lemus et al. 2018). The overarching pathophysiology of MS is outlined in Fig. 1.2.

1.5 MS from a Genetic Perspective

The heritability of MS has been extensively explored, and it was found that 15% of MS patients have affected family members, with a disease risk approaching 1 in 25 if a sibling having the disease (Goris et al. 2022; Kasper et al. 2018). Genome-wide association studies (GWAS) concluded that MS heritability is governed by many genetic factors rather than a single, isolated mutation, with the majority of those genes associated with factors that play a role in the adaptive immunity (Goris et al. 2022). One of the earliest MS genetic risk factors identified is the HLA-DRB1*15:01, which increases the risk of the disease by threefold. HLA-DRB1 is expressed as antigen-presenting proteins on antigen-presenting cells (APC) that specifically present myelin and EBV peptides (Parnell and Booth 2017). Other recently recognized genes are those that translate into the IL-7 receptor (CD127), IL-2 receptor (CD25), T cell costimulatory molecule LFA-3 (CD58), and tyrosine kinase 2 (TYK2). Additional genes implicated in MS include EOMES and TBX21 encoding eomesodermin and T-bet, respectively. These genes are transcription factors regulating natural killers, $CD8^+$ memory cells, and $CD4^+$ differentiation. (Couturier et al. 2011; Goris et al. 2022; Kasper et al. 2018; McKay et al. 2016; Parnell and Booth 2017).

1.6 Environmental Factors

Research suggests that the development and progression of MS are influenced by a multifaceted interplay between genetic predisposition and distinct environmental factors, collectively elevating susceptibility. In this section, we will provide a concise overview of various environmental factors based on existing literature, with a more comprehensive exploration in Chap. 2 of the book. Broadly, studies have identified environmental risk factors that interact with human leukocyte antigen (HLA) risk genes and may play a role in the development or exacerbation of MS. These factors include inadequate sun exposure and low vitamin D levels, obesity, smoking, and infection with the Epstein-Barr virus. For a more detailed exploration, readers are encouraged to refer to Chap. 2 of this book.

While a comprehensive discussion of the role of vitamin D is addressed in Chap. 8, we will briefly touch upon key aspects in this section. Vitamin D deficiency and reduced sunlight exposure have long been associated with an elevated risk of MS, prompting investigations into the potential impact of vitamin D supplementation on disease progression. Having anti-inflammatory properties, vitamin D plays a crucial role as an immunomodulator and potentially reducing inflammation in MS. It also plays a vital role in bone metabolism. Notably, its synthesis primarily relies on exposure to ultraviolet (UV) radiation from sunlight. Numerous studies have assessed vitamin D levels in the MS population, revealing correlations between low serum vitamin D levels and an increased risk of MS (Michel 2018; Touil et al. 2023). Some research suggests that vitamin D deficiency in MS patients could stem from issues with the vitamin D receptor (VDR), which is essential for activating the VDR pathway, thereby enabling vitamin D to initiate an anti-inflammatory cascade (Touil et al. 2023). Adequate vitamin D levels support neuronal health by affecting neurotrophic factor secretion and enhancing neuronal survival (Touil et al. 2023). Vitamin D supplementation and appropriate sun exposure hold promise in managing MS risk and potentially impacting disease progression. Several randomized control trials (RCTs) investigating the effect of vitamin D supplementation found that they were only beneficial when patients had an existing vitamin D deficiency. Unfortunately, investigations on the impact of vitamin D on MS are limited by their small sample sizes and short durations. Additional research is needed to elucidate the utility of vitamin D and its implication on the pathophysiology of MS (Waubant et al. 2019).

Obesity has gained recognition as a significant risk factor in the pathophysiology of MS. Observational studies consistently demonstrate an approximate two-fold increased risk of developing pediatric and adult MS in individuals with obesity since adolescence and early adulthood, compared to individuals with normal weight (Alfredsson and Olsson 2019; Waubant et al. 2019). One crucial mechanism linking obesity and MS progression is inflammation. Obesity triggers inflammation by elevating levels of IL-17 and IL-23, thereby promoting the presence of pro-inflammatory Th17 T cells. Furthermore, obesity has been associated with elevated levels of the inflammatory mediator leptin, primarily produced by adipose tissue, which influences immune responses and can exacerbate inflammation in individuals with obesity. Studies have also demonstrated that increased body weight is related to decreased levels of vitamin D availability. Vitamin D deficiency in individuals with obesity can exacerbate inflammation in the body, potentially worsening the course of MS (Touil et al. 2023). Cumulatively, chronic inflammation associated with obesity could potentially impact the progression of MS, which makes addressing it part of the comprehensive management of MS patients.

Cigarette smoke is a well-established risk factor for MS, contributing to the onset and exacerbation of the disease. Even the slightest lung irritation through secondhand smoke is associated with a comparable increased risk for MS as first-hand smoking. This increased risk is primarily attributed to the irritation of lung tissues (Alfredsson and Olsson 2019; Waubant et al. 2019). Some research proposes that smoking may activate immune cells near the airways or may trigger the aryl hydrocarbon receptor, known to influence the immune response (Waubant et al. 2019). Alternative mechanisms have been suggested where smoking has direct effects on the CNS, including the production of NO which can lead to axonal damage, and the potential to directly affect the permeability of the BBB (Correale et al. 2013). The evidence points to a clear and proportional relationship between smoking and the risk of developing MS (Alfredsson and Olsson 2019; Waubant et al. 2019). A study conducted in 2001 involving over 100,000 American women in the Nurses' Health Study (NHS) and NHS II underscored the association between active smoking and a 1.6 times higher incidence of MS compared to non-smokers. This risk escalates with cumulative exposure to tobacco. Additionally, it has been observed that sustained smoking can expedite the progression of MS, particularly toward SPMS, as well as increase disease severity and accelerated progression to disability (Michel 2018). What is intriguing is that smoking heightens the risk of MS regardless of the age at which one is exposed, with both the duration and intensity of smoking contributing independently to this risk (Correale et al. 2013; Michel 2018). Furthermore, individuals who carry the major MS risk allele HLA-DRB1*15:01, without the protective HLA-A2 variant, have an elevated risk of MS. This risk is further magnified in the presence of smoking, underscoring the intricate interplay between genetic susceptibility and smoking in the context of MS (Alfredsson and Olsson 2019; Michel 2018). Importantly, studies have shown that this risk of MS associated with smoking is reversible, such that if a smoker quits smoking for approximately 10 years, their risk of developing MS can return to a level comparable to that of non-smokers (Michel 2018; Waubant et al. 2019), thus making smoking cessation a promising therapeutic avenue in MS.

1.7 Managing MS

Managing MS requires a multifaceted approach. This includes managing symptoms, providing physiotherapy, and addressing MS relapses. The treatment strategies for MS typically encompass two main approaches. The first approach is a 'gradual escalation' strategy, where treatment begins with a relatively mild or less potent therapy and is adjusted, as needed, based on the patient's response and disease progression. The second approach is a 'higher initial intensity' strategy, which involves starting treatment with a more potent therapy from the outset, with the possibility of reducing its intensity if the patient's condition stabilizes (Travers et al. 2022). For a comprehensive understanding of the current treatments and their implications, a detailed explanation will be available in Chap. 16 of this book.

At the core of MS treatment are disease-modifying therapies (DMTs), which play a pivotal role by reducing relapses and slowing disease progression (Charabati et al. 2023; Travers et al. 2022). This class of medications are categorized into three subgroups: Personalized Medicine, Novel Mechanisms of Action, and Emerging Oral and Infusion Therapies (Robertson and Moreo 2016). Ongoing clinical trials aim to improve the effectiveness, safety, and overall impact of these therapies on the quality of life for individuals with MS (Claflin et al. 2018).

Personalized Medicine tailors treatment based on individual attributes, offering a patient-centered, customized approach to treatment. Tailored treatments can be synthesized through an understanding of the patient's genetic profile, biomarkers, personal preferences, comorbidities, and adherence to therapy (Giovannoni 2017). Novel Mechanisms of Action explore various avenues for targeting the immune system, including therapies that focus on B cells, CD20, sphingosine 1 phosphate (S1P), and alpha-4 integrin interactions (Cross and Naismith 2014; Subei and Cohen 2015).

Emerging oral and infusion therapies, exemplified by monoclonal antibodies, like Ocrelizumab, Ofatumumab, Ublituximab, and immunomodulators, namely Fingolimod, offer promising approaches for managing MS. These treatments selectively target various immune system components to reduce relapses and impede disease progression. Some therapies combat inflammation and impede demyelination, while others focus on restraining lymphocyte activity, preventing their detrimental infiltration into the central nervous system, thereby alleviating inflammation. Additionally, certain treatments modify immune cell trafficking, collectively working towards more effective MS management (Castro-Borrero et al. 2012; Cross and Naismith 2014; Subei and Cohen 2015).

1.8 Multiple Sclerosis and the Role of Diet

Dietary components and regimens are associated with the progression, relapse rate, and development of MS. Several studies investigated the various regimens and supplements that can modulate the pathogenesis of MS both positively and negatively. Vitamins, such as A and D, and dietary supplements, including caffeine and carnitine, can benefit MS patients (Jakimovski et al. 2019; Nunes and Piuvezam 2019; Tryfonos et al. 2019). Additionally, the literature emphasizes the importance of vitamin D in lowering the risk of developing MS. Vitamin D has been implicated in several studies as an immunomodulator with regulatory effects on the inflammatory processes of MS (Chang et al. 2010; Muthian et al. 2006; Staeva-Vieira and Freedman 2002). In fact, lower vitamin D serum levels have been associated with a poor prognosis (Tryfonos et al. 2019).

Similarly, diets such as the Western or high salt diet (HSD) can negatively impact the prognosis of MS patients. A regimen of dietary components high in salt, saturated fats, and carbohydrates damages the gut microbiome, increasing intestinal permeability and triggering an escalated inflammatory response that promotes demyelination of the CNS (Jayasinghe et al. 2022). The chapters in this book will shed light on the potential role of diet in the development and progression of MS, discuss some of the mechanistic pathways involved in regulating its pathogenesis, and consider the effects of dietary components and regimens on the inflammatory markers linked to MS.

1.9 Summary

This chapter explores MS a chronic central nervous system disease characterized by autoimmune inflammation, demyelination, and axonal degeneration, primarily affecting young adults, especially women. MS presents diverse symptoms like sensory deficits, motor issues, and optic neuritis. Diagnosis involves clinical signs, MRI, cerebrospinal fluid analysis, and evoked potentials. It is classified into relapsing-remitting, secondary progressive, primary progressive, and progressive relapsing subtypes. MS pathophysiology involves immune cells, microglia activation, and inflammatory mediators. Genetic factors, particularly the HLA-DRB1*15:01 allele, and environmental factors such as vitamin D deficiency, obesity, smoking, and Epstein-Barr virus infection contribute to its development. Management includes symptom control, physiotherapy, and DMTs. The chapter also highlights diet as a potential preventive and therapeutic target. Vitamins A and D, caffeine, and carnitine can benefit MS patients, while high-salt, high-fat diets negatively impact the disease. The book emphasizes the potential of dietary interventions in managing MS across various cultural and socioeconomic backgrounds.

References

Alfredsson L, Olsson T (2019) Lifestyle and environmental factors in multiple sclerosis. Cold Spring Harb Perspect Med 9(4):a028944. https://doi.org/10.1101/cshperspect.a028944

Castro-Borrero W, Graves D, Frohman TC, Flores AB, Hardeman P, Logan D, Orchard M, Greenberg B, Frohman EM (2012) Current and emerging therapies in multiple sclerosis: a systematic review. Ther Adv Neurol Disord 5(4):205–220. https://doi.org/10.1177/1756285612450936

Chang J-H, Cha H-R, Lee D-S, Seo KY, Kweon M-N (2010) 1,25-Dihydroxyvitamin D3 inhibits the differentiation and migration of TH17 cells to protect against experimental autoimmune encephalomyelitis. PLoS One 5(9):e12925. https://doi.org/10.1371/journal.pone.0012925

Charabati M, Wheeler MA, Weiner HL, Quintana FJ (2023) Multiple sclerosis: neuroimmune crosstalk and therapeutic targeting. Cell 186(7):1309–1327. https://doi.org/10.1016/j.cell.2023.03.008

Choi IY, Piccio L, Childress P, Bollman B, Ghosh A, Brandhorst S, Suarez J, Michalsen A, Cross AH, Morgan TE, Wei M, Paul F, Bock M, Longo VD (2016) A diet mimicking fasting promotes regeneration and reduces autoimmunity and Multiple sclerosis symptoms. Cell Rep 15(10):2136–2146. https://doi.org/10.1016/j.celrep.2016.05.009

Ciccarelli O, Barkhof F, Bodini B, De Stefano N, Golay X, Nicolay K, Pelletier D, Pouwels PJ, Smith SA, Wheeler-Kingshott CA, Stankoff B, Yousry T, Miller DH (2014) Pathogenesis of Multiple sclerosis: insights from molecular and metabolic imaging. Lancet Neurol 13(8):807–822. https://doi.org/10.1016/S1474-4422(14)70101-2

Claflin SB, Broadley S, Taylor BV (2018) The effect of disease modifying therapies on disability progression in Multiple sclerosis: a systematic overview of meta-analyses. Front Neurol 9:1150. https://doi.org/10.3389/fneur.2018.01150

Correale J, Balbuena Aguirre ME, Farez MF (2013) Sex-specific environmental influences affecting MS development. Clin Immunol 149(2):176–181. https://doi.org/10.1016/j.clim.2013.02.006

Couturier N, Bucciarelli F, Nurtdinov RN, Debouverie M, Lebrun-Frenay C, Defer G, Moreau T, Confavreux C, Vukusic S, Cournu-Rebeix I, Goertsches RH, Zettl UK, Comabella M, Montalban X, Rieckmann P, Weber F, Müller-Myhsok B, Edan G, Fontaine B, Brassat D (2011) Tyrosine kinase 2 variant influences T lymphocyte polarization and Multiple sclerosis susceptibility. Brain 134(3):693–703. https://doi.org/10.1093/brain/awr010

Cross AH, Naismith RT (2014) Established and novel disease-modifying treatments in Multiple sclerosis. J Intern Med 275(4):350–363. https://doi.org/10.1111/joim.12203

Doshi A, Chataway J (2016) Multiple sclerosis, a treatable disease. Clin Med (Lond) 16(Suppl 6):s53–s59. https://doi.org/10.7861/clinmedicine.16-6-s53

Ford H (2020) Clinical presentation and diagnosis of Multiple sclerosis. Clin Med (Lond) 20(4):380–383. https://doi.org/10.7861/clinmed.2020-0292

Giovannoni G (2017) Personalized medicine in Multiple sclerosis. Neurodegener Dis Manag 7(6s):13–17. https://doi.org/10.2217/nmt-2017-0035

Goris A, Vandebergh M, McCauley JL, Saarela J, Cotsapas C (2022) Genetics of Multiple sclerosis: lessons from polygenicity. Lancet Neurol 21(9):830–842. https://doi.org/10.1016/S1474-4422(22)00255-1

Jakimovski D, Guan Y, Ramanathan M, Weinstock-Guttman B, Zivadinov R (2019) Lifestyle-based modifiable risk factors in Multiple sclerosis: review of experimental and clinical findings. Neurodegener Dis Manag 9(3):149–172. https://doi.org/10.2217/nmt-2018-0046

Jayasinghe M, Prathiraja O, Kayani AMA, Jena R, Caldera D, Silva MS, Singhal M, Pierre J (2022) The role of diet and gut microbiome in Multiple sclerosis. Cureus 14(9):e28975. https://doi.org/10.7759/cureus.28975

Kasper DL, Fauci AS, Hauser SL, Longo DL, Jameson JL, Loscal J (2018) Multiple sclerosis. In: Harrison's principles of internal medicine, vol 1 and 2. McGraw-Hill Education, New York, pp 3188–3200

Lemus HN, Warrington AE, Rodriguez M (2018) Multiple sclerosis: mechanisms of disease and strategies for myelin and axonal repair. Neurol Clin 36(1):1–11. https://doi.org/10.1016/j.ncl.2017.08.002

Lublin FD, Reingold SC, Cohen JA, Cutter GR, Sorensen PS, Thompson AJ, Wolinsky JS, Balcer LJ, Banwell B, Barkhof F, Bebo B Jr, Calabresi PA, Clanet M, Comi G, Fox RJ, Freedman MS, Goodman AD, Inglese M, Kappos L, Polman CH (2014) Defining the clinical course of Multiple sclerosis: the 2013 revisions. Neurology 83(3):278–286. https://doi.org/10.1212/WNL.0000000000000560

Lucchinetti C, Bruck W, Parisi J, Scheithauer B, Rodriguez M, Lassmann H (2000) Heterogeneity of Multiple sclerosis lesions: implications for the pathogenesis of demyelination. Ann Neurol 47(6):707–717. https://doi.org/10.1002/1531-8249(200006)47:6<707::aid-ana3>3.0.co;2-q

McKay FC, Gatt PN, Fewings N, Parnell GP, Schibeci SD, Basuki MA, Powell JE, Goldinger A, Fabis-Pedrini MJ, Kermode AG, Burke T, Vucic S, Stewart GJ, Booth DR (2016) The low EOMES/TBX21 molecular phenotype in Multiple sclerosis reflects CD56+ cell dysregulation and is affected by immunomodulatory therapies. Clin Immunol 163:96–107. https://doi.org/10.1016/j.clim.2015.12.015

Michel L (2018) Environmental factors in the development of Multiple sclerosis. Rev Neurol (Paris) 174(6):372–377. https://doi.org/10.1016/j.neurol.2018.03.010

Muthian G, Raikwar HP, Rajasingh J, Bright JJ (2006) 1,25 dihydroxyvitamin-D3 modulates JAK–STAT pathway in IL-12/IFNγ axis leading to Th1 response in experimental allergic encephalomyelitis. J Neurosci Res 83(7):1299–1309. https://doi.org/10.1002/jnr.20826

Nij Bijvank JA, van Rijn LJ, Balk LJ, Tan HS, Uitdehaag BMJ, Petzold A (2019) Diagnosing and quantifying a common deficit in Multiple sclerosis: Internuclear ophthalmoplegia. Neurology 92(20):e2299–e2308. https://doi.org/10.1212/wnl.0000000000007499

Nunes ACDF, Piuvezam G (2019) Nutritional supplementation of vitamin a and health-related outcomes in patients with Multiple sclerosis: a protocol for a systematic review and meta-analysis of randomized clinical trials. Medicine 98(25):e16043. https://doi.org/10.1097/md.0000000000016043

Opara JA, Brola W, Wylegala AA, Wylegala E (2016) Uhthoff's phenomenon 125 years later—what do we know today? J Med Life 9(1):101–105

Parnell GP, Booth DR (2017) The Multiple sclerosis (MS) genetic risk factors indicate both acquired and innate immune cell subsets contribute to MS pathogenesis and identify novel therapeutic opportunities [mini review]. Front Immunol 8:425. https://doi.org/10.3389/fimmu.2017.00425

Petzold A, Wattjes MP, Costello F, Flores-Rivera J, Fraser CL, Fujihara K, Leavitt J, Marignier R, Paul F, Schippling S, Sindic C, Villoslada P, Weinshenker B, Plant GT (2014) The investigation of acute optic neuritis: a review and proposed protocol. Nat Rev Neurol 10(8):447–458. https://doi.org/10.1038/nrneurol.2014.108

Polman CH, Reingold SC, Banwell B, Clanet M, Cohen JA, Filippi M, Fujihara K, Havrdova E, Hutchinson M, Kappos L, Lublin FD, Montalban X, O'Connor P, Sandberg-Wollheim M, Thompson AJ, Waubant E, Weinshenker B, Wolinsky JS (2011) Diagnostic criteria for Multiple sclerosis: 2010 revisions to the McDonald criteria. Ann Neurol 69(2):292–302. https://doi.org/10.1002/ana.22366

Robertson D, Moreo N (2016) Disease-modifying therapies in Multiple sclerosis: overview and treatment considerations. Fed Pract 33(6):28–34. https://www.ncbi.nlm.nih.gov/pubmed/30766181

Staeva-Vieira TP, Freedman LP (2002) 1,25-Dihydroxyvitamin D3 inhibits IFN-γ and IL-4 levels during in vitro polarization of primary murine CD4+ T Cells1. J Immunol 168(3):1181–1189. https://doi.org/10.4049/jimmunol.168.3.1181

Subei AM, Cohen JA (2015) Sphingosine 1-phosphate receptor modulators in Multiple sclerosis. CNS Drugs 29(7):565–575. https://doi.org/10.1007/s40263-015-0261-z

Touil H, Mounts K, De Jager PL (2023) Differential impact of environmental factors on systemic and localized autoimmunity. Front Immunol 14:1147447. https://doi.org/10.3389/fimmu.2023.1147447

Travers BS, Tsang BK, Barton JL (2022) Multiple sclerosis: diagnosis, disease-modifying therapy and prognosis. Aust J Gen Pract 51(4):199–206. https://doi.org/10.31128/AJGP-07-21-6103

Tryfonos C, Mantzorou M, Fotiou D, Vrizas M, Vadikolias K, Pavlidou E, Giaginis C (2019) Dietary supplements on controlling Multiple sclerosis symptoms and relapses: current clinical evidence and future perspectives. Medicines 6(3):95. https://www.mdpi.com/2305-6320/6/3/95

Waubant E, Lucas R, Mowry E, Graves J, Olsson T, Alfredsson L, Langer-Gould A (2019) Environmental and genetic risk factors for MS: an integrated review. Ann Clin Transl Neurol 6(9):1905–1922. https://doi.org/10.1002/acn3.50862

Wootla B, Eriguchi M, Rodriguez M (2012) Is Multiple sclerosis an autoimmune disease? Autoimmune Dis 2012:969657. https://doi.org/10.1155/2012/969657

Chapter 2
Life Chapters: Navigating Multiple Sclerosis Across Pregnancy, Breastfeeding, Epidemiology, and Beyond

Salsabil Zubedi, Hana Al-Ali, Nadia Rabeh, Sara Aljoudi, Zakia Dimassi, and Hamdan Hamdan

Abstract Multiple sclerosis (MS) entails a complex interplay of various factors that significantly impact the progression of the disease. This chapter covers the multifaceted dynamics of MS during pregnancy, examining fertility, the impact of pregnancy on MS, and the long-term progression. Despite no evidence of reduced fertility in individuals with MS, concerns about disease transmission and lower childbearing rates persist, leading to a lower average number of children per woman. Pregnancy, a natural modifier for MS, is associated with a significant decrease in relapse rates during the third trimester. However, the postpartum phase poses an elevated risk of relapses, attributed to the sudden removal of protective factors during pregnancy. Multiple pregnancies diminish the likelihood of developing MS, while the impact of consecutive pregnancies on disease progression remains complex. The review also discusses obstetric outcomes, disease-modifying therapies (DMTs), and the influence of various components such as body mass index (BMI), Epstein–Barr virus (EBV), vitamin D deficiency, and smoking on MS susceptibility. The complex interplay of genetic predisposition and environmental elements in shaping the risk and progression of MS is emphasized, providing valuable insights for clinicians and researchers navigating family planning and MS management.

S. Zubedi · H. Al-Ali · N. Rabeh · S. Aljoudi · H. Hamdan (✉)
Department of Biological Sciences, College of Medicine and Health Sciences, Khalifa University, Abu Dhabi, United Arab Emirates
e-mail: hamdan.hamdan@ku.ac.ae

Z. Dimassi
Department of Medical Sciences, College of Medicine and Health Sciences, Khalifa University, Abu Dhabi, United Arab Emirates

© The Author(s), under exclusive license to Springer Nature Singapore Pte Ltd. 2024
H. Hamdan (ed.), *Exploring the Effects of Diet on the Development and Prognosis of Multiple Sclerosis (MS)*, Nutritional Neurosciences,
https://doi.org/10.1007/978-981-97-4673-6_2

Keywords Multiple sclerosis (MS) · Pregnancy · Breastfeeding · Environmental factors · Obesity · Smoking

Abbreviations

AFC	Antral follicle count
AMH	Anti-Mullerian hormone
ARR	Annualized relapse rate
BBB	Blood–brain barrier
BMI	Body mass index
CIS	Clinical isolated syndrome
CNS	Central nervous system
DMTs	Disease-modifying therapies
EAE	Experimental autoimmune encephalomyelitis
EBV	Epstein–Barr virus
EDSS	Expanded disability status scale
FoxP3	Forkhead box P3
FSH	Follicle-stimulating hormone
GA	Glatiramer acetate
HHV-4	Human herpesvirus 4
HLA	Human leukocyte antigen
IFN-β	Interferon-beta
IL-17	Interleukin-17
LH	Luteinizing hormone
MR	Mendelian randomization
MRI	Magnetic resonance imaging
MS	Multiple sclerosis
NHS	Nurses' Health Study
NK	Natural killer
NO	Nitric oxide
OCP	Oral contraceptives
OR	Odds ratio
PPMS	Primary progressive MS
PRIMS	Pregnancy-related relapse in MS
RRMS	Relapsing-remitting MS
SAT	Subcutaneous adipose tissue
SPMS	Secondary-progressive MS
T2-LV	T2 lesion volume
(TGF)-β	Transforming growth factor-β
(Th2) cells	T helper 2
TNF-α	Tumor necrosis factor-α
Tregs	Regulatory T cells
UV	Ultraviolet

VAT Visceral adipose tissue
VDR Vitamin D receptor

Learning Objectives
- Learn about how pregnancy affects MS, including the significant reduction in relapse rates during the third trimester and the increased risk of relapses postpartum.
- Understand the evidence regarding fertility levels in women with MS and the factors contributing to lower childbearing rates.
- Investigate the relationship between multiple pregnancies and the risk of developing MS.
- Evaluate the role of disease-modifying therapies (DMTs) during pregnancy and breastfeeding.
- Identify environmental and lifestyle factors affecting MS susceptibility and progression.

2.1 Introduction

Within the confines of this chapter, a comprehensive exploration unfolds, meticulously investigating the intricate interplay of environmental and individual factors influencing the landscape of multiple sclerosis (MS). Research has supported the complex interaction between MS and environmental and individual factors (Hellwig et al. 2015; Mokry et al. 2016; Sangha et al. 2023; Vandebergh et al. 2022; Wallin et al. 2019). Of the factors studied, obstetric outcomes, pregnancy, and breast feeding have been intensively studied. These factors, in relation to existing disease-modifying therapies (DMTs), are also discussed, as it is currently unclear whether DMTs pose serious ramifications to breastfeeding mothers and their newborn's development. The exposition of the multifaceted dynamics of MS during pregnancy is presented, encompassing fertility considerations, the modulatory effects of pregnancy on MS, and the enduring ramifications for long-term progression. Insights into the protective influence of pregnancy on MS, the consequences of consecutive pregnancies on disease progression, and the application of outcome measures to assess the status of individuals with MS are methodically delineated (Carbone et al. 2023; Dalla Costa et al. 2019; Haas and Hommes 2007; Hellwig et al. 2015; Sicotte et al. 2002).

Furthermore, the discourse scrutinizes environmental determinants, spanning from insufficient sun exposure and low vitamin D levels to the intricate influence of obesity. The potential implications of the Epstein–Barr virus (EBV) in MS development and progression are examined, and the association between body mass index (BMI) and MS is expounded, with a focus on the involvement of inflammatory pathways and mediators (Harroud et al. 2021; Mokry et al. 2016; Mortazavi et al.

2023; Sangha et al. 2023; Waubant et al. 2019). Transitioning to the forefront of research methodologies, this chapter explores Mendelian randomization (MR) studies, shedding light on the nuanced relationship between BMI and the risk of developing MS. The narrative seamlessly transitions into a research paper, scrutinizing the complex interrelation between MS risk and obesity measures, including BMI and visceral adipose tissue (VAT). These findings underscore potential shared biological pathways involving abnormal immune responses and inflammatory processes (Vandebergh et al. 2022). The discourse extends to the established association between cigarette smoking and MS initiation and exacerbation. A scientific canvas is painted, illustrating oxidative stress, immune cell activation, and autoimmune response modulation as key elements contributing to the heightened risk of MS associated with smoking (Marabita et al. 2017; Wingerchuk 2012). Temporal considerations, approximately 10 years post-cessation, are discussed, revealing a diminishing risk, yet smoking's enduring influence on progressive disease and accelerated disability progression is examined (Michel 2018; Waubant et al. 2019).

Geographical considerations emerge as crucial factors in the epidemiological context of MS, showcasing significant variations in prevalence and incidence rates (Wallin et al. 2019). Higher latitudes are linked with elevated MS prevalence, and regions of higher socioeconomic status bear a burden of increased MS prevalence (Wallin et al. 2019). This chapter concludes with an exploration of environmental factors, including pollution, sunlight exposure, and smoking, as potential determinants influencing the degree of MS burden in specific regions. In essence, this chapter offers a scholarly journey, presenting a meticulous tapestry of insights into the environmental and individual facets influencing the narrative of multiple sclerosis.

2.2 Pregnancy

2.2.1 Fertility

Up to the present time, the existing evidence indicates that, when compared to healthy controls, individuals with MS do not experience reduced fertility and can successfully completing a pregnancy—compared to their healthy counterparts (Voskuhl and Momtazee 2017). Consequently, fertility levels in women with MS exhibit no significant variation between periods prior to and following the diagnosis of MS, and the rates of miscarriage align with those observed in the general public. Nevertheless, compared to the general population, women with MS tend to bear, on average, fewer children. This discrepancy could be attributed to concerns about the children inheriting the condition, particularly if both parents have the condition or if there is a family history. This concern arises because the likelihood of a child developing MS is 2% when one parent has MS and increases to 6–12% when both parents are affected. Other contributing factors to this observation may involve sexual dysfunction resulting from the condition or the medical regimen's cytotoxic impact, particularly in cases of progressive MS (Alhomoud et al. 2021).

Carbone et al. (2023) compared markers associated with ovarian reserve between women diagnosed with MS and a control group of healthy women using levels of anti-Mullerian hormone (AMH) in serum circulation. AMH plays a role in the quantity of developing follicles and their readiness for ovulation. No notable differences were found in the levels of AMH in serum circulation between women diagnosed with MS and the control group. In contrast, women affected by MS demonstrated significantly lower antral follicle count (AFC), an indicator for egg supply for future use (ovarian reserve) during the early follicular phase, and lower serum estradiol levels compared to their healthy counterparts. Conversely, levels of luteinizing hormone (LH), a hormone that aids in ovulation, were notably higher in women with MS than in the control group. Meanwhile, no discernible differences were observed in the mean follicle-stimulating hormone (FSH) and ovarian volume levels between women with MS and the control group (Carbone et al. 2023).

2.2.2 Effect of Pregnancy on MS

Just a few decades ago, women diagnosed with MS were discouraged from pursuing pregnancy due to limited data on the potential effects of pregnancy on their MS journey and concerns about possible complications related to pregnancy and childbirth. In 1998, the pregnancy-related relapse in MS (PRIMS) study marked a pivotal moment as the first significant and prospective study examining MS relapses during pregnancy. The study revealed that MS does not directly influence the rates of miscarriage, stillborn infants, congenital fetal defects, or ectopic pregnancy (Confavreux et al. 1998). Notably, the study demonstrated that the incidence of these complications was comparable to that observed in the general population. Additional research corroborated these findings, emphasizing that MS does not adversely affect fertility or the occurrence of fetal abnormalities (Stuart and Bergstrom 2011).

MS, a chronic autoimmune condition, exhibits a susceptibility and course that are influenced by reproductive factors (Langer-Gould et al. 2020). The prevalence of MS is on the rise, particularly in women, with a women-to-men ratio reaching as high as 3.2:1, especially during the reproductive years, thereby elevating concerns about pregnancy and childbirth in women with MS (Alhomoud et al. 2021). Research shows that the third trimester of pregnancy acted as a natural modifier of MS, as relapse rates decreased by 70% (Voskuhl and Momtazee 2017). Substantial evidence, including systematic reviews and a national consensus, has collectively determined that pregnancy exerts a protective influence on MS; thus, it is inaccurate to classify pregnant women with MS as high risk (Alhomoud et al. 2021). The decrease in relapse rates during the final trimester, potentially the result of the anti-inflammatory effect of hormonal changes (increased levels of estrogens, progesterone, prolactin, etc.), is regarded as the most effective disease-modifying treatment for MS (Voskuhl and Momtazee 2017). Nevertheless, this observed effect has no bearing on long-term disease progression and postpartum relapse is bound to occur in the first 3 postpartum months (Vukusic et al. 2021).

There are three primary outcome measures used to evaluate the status of individuals with MS. (1) Annualized relapse rate (ARR) denotes the average number of relapses experienced by a group of patients in a clinical study within a year; (2) expanded disability status scale (EDSS) involves a neurological examination where the score out of 10 corresponds to the level of disability, with higher scores indicating increasingly severe disability; and (3) biological markers, specifically magnetic resonance imaging (MRI), which aid in examining lesions. Pregnancy introduces numerous changes that may contribute to immunomodulation, including a rise in estrogen levels (estradiol and estriol) and progesterone, among others. These levels steadily increase during pregnancy, reaching their maximum in the final trimester, coinciding with the period of heightened disease protection (Voskuhl and Momtazee 2017). Preclinical investigations demonstrated that estriol, a pregnancy estrogen, conferred protective effects in experimental autoimmune encephalomyelitis (EAE) when administered at a concentration aligning with physiological levels during pregnancy (Voskuhl and Momtazee 2017). In addition to estrogen levels reaching maximum levels at later stages of pregnancy, they have also exhibited neuroprotective properties in various neurological disease models, playing a recognized part in regular cognitive development (Voskuhl and Momtazee 2017).

A clinical study was conducted to assess the therapeutic potential of estriol in nonpregnant female patients with relapsing-remitting MS (RRMS) (Sicotte et al. 2002). These patients received oral estriol at a daily dose of 8 mg. The results demonstrated notable improvements from the pretreatment baseline, including lower IFN-γ levels in mononuclear cells found in the peripheral blood, reduced delayed-type hypersensitivity responses to tetanus, and diminished numbers and volumes of gadolinium-enhancing lesions on cerebral MRI every month. During the initial 3 months of treatment, RRMS patients experienced a 79% reduction in median total enhancing lesion volumes ($p = 0.02$) and an 82% decrease in lesion numbers ($p = 0.09$), and these reductions were sustained in the subsequent 3 months, with volumes decreasing by 82% ($p = 0.01$) and numbers by 82% ($p = 0.02$) (Sicotte et al. 2002). Importantly, the beneficial effects were reversible, as discontinuation of estriol treatment led to an increase in enhancing lesions to levels seen before the treatment, and reintroducing estriol resulted in a significant decrease in enhancing lesions. Conversely, the administration of high levels of progestin derivative, nomegestrol acetate, and reduced concentrations of 17-beta-estradiol promptly after delivery to induce hormonal pregnancy impregnation on postpartum relapse rates was studied in a European multicenter placebo-controlled and randomized phase 3 study (POPARTMUS). Unfortunately, the results of the studied showed that the therapy was unsuccessful in preventing postpartum relapse (Vukusic et al. 2021). Another trial involving estradiol showed a dose-dependent trend of benefit, whereas a trial utilizing progesterone treatment yielded negative results (Voskuhl and Momtazee 2017).

In another clinical trial, patients were treated with synthetic estradiol, a component found in oral contraceptives (OCP), administered at concentrations designed to achieve serum estradiol levels consistent with those in the natural ovarian cycle.

However, it was found that using high estradiol doses to replicate estradiol levels in the last trimester of pregnancy was deemed impractical due to potential adverse effects, as it strongly binds to the estrogen receptor alpha found in uterine and breast tissue (Voskuhl and Momtazee 2017). Notably, a study conducted by Ghajarzadeh et al. (2022) found no association between the consumption of OCP and an increased risk of developing MS. Additionally, there are no contraindications for the use of hormonal contraception, and estrogen is believed to have protective effects against the onset of MS (Stuart and Bergstrom 2011).

2.2.3 Relapse Rate: Postpartum Period

Unlike the protective effects observed during pregnancy, the postpartum phase is associated with an elevated risk of relapses, potentially attributed to the sudden removal of protective factors during pregnancy (Voskuhl and Momtazee 2017). Evidence from the literature indicated a decline in the relapse rate toward the late stages of pregnancy, followed by a temporary increase in the initial 6 months after childbirth (Holmøy and Torkildsen 2016).

A history of active relapsing MS before pregnancy is a strong and reliable indicator that a patient will endure a postpartum relapse (Voskuhl and Momtazee 2017). The transient surge in relapses diminishes following the first postpartum year, with rates reverting to the baseline levels before pregnancy (Voskuhl and Momtazee 2017). In one of the most extensive international studies examining the effect of pregnancy on MS relapse rates, data from the global MSBase Registry were utilized (Hughes et al. 2014). With data on 893 pregnancies in 674 women with MS, the results demonstrated that relapse rates were lower than those observed in the PRIMS study by at least 50% with no alterations in EDSS. Comparing the preconception period to the last trimester, the ARR declined from 0.3 to 0.1, followed by an increase to 0.6 in the early postpartum period. A notable predictor for postpartum relapse was the ARR before conception. Additionally, the study showed a 45% reduction in postpartum relapse when disease-modifying therapies (DMTs) were taken before conception (Alhomoud et al. 2021; Hughes et al. 2014).

During pregnancy, there may be an increase in T helper 2 (Th2) cells, known for their anti-inflammatory effects, and the release of interleukin-10 cytokines by the placenta, as opposed to the pro-inflammatory effects associated with Th1 cells. In the last trimester of pregnancy, a decline in the percentage of circulating Natural Killer (NK) cells has been documented (Saraste et al. 2007). The elevated levels of sex hormones during pregnancy contribute to immune modulation and neuroprotective effects. The subsequent rise in MS disease activity after childbirth can be attributed to the modest state of humoral immunity and decreased interleukin levels (Alhomoud et al. 2021). This phenomenon is thought to be linked to immunological and hormonal shifts in pregnancy. Local immunosuppression typically occurs to protect the fetus during pregnancy, characterized by a shift in T helper cells from predominantly Th1 to more of a Th2 type occurs. NK cell cytolytic activity tends to decrease during pregnancy, particularly in the third trimester, and returns to prepregnancy levels during the postpartum period (Stuart and Bergstrom 2011).

Predicting relapses during pregnancy and the postpartum phase can be done by assessing the level of disease activity both prior to and during pregnancy, along with the use of potent DMTs before conception. What was found to be crucial in lowering the incidence of postpartum relapses was the timely reintroduction of DMTs (Alhomoud et al. 2021). The research conducted by Langer-Gould et al. 2020, indicated that a heightened frequency of relapses before pregnancy correlated with relapses during the entire postpartum year, excluding the first 6 months.

A meta-analysis encompassing 13 studies by Finkelsztejn et al. (2011) revealed that in the year before pregnancy, yearly relapse rates averaged 0.435 ± 0.021; these rates diminished to 0.182 ± 0.012 or 0 during pregnancy, then re-increased to 0.703 ± 0.024 or 0.758 postpartum. The duration for assessing postnatal relapse rates varied from 3 to 12 months among different authors. A notable reduction in relapse activity during pregnancy was observed despite the level of disability remaining constant during both pregnancy and the postpartum period. Contrarily, a slightly higher incidence of relapses was observed after childbirth compared to the preconception period (McKay et al. 2015).

Compared to pre- and postpartum periods, the percentage of women with MRI lesion activity was similar (McKay et al. 2015). In the postpartum phase, women exhibited increased T2 lesion volume (T2-LV) and a higher T2-LV increase compared to the pregnancy period. Since the atrophied T2-LV is a result of inflammation in the form of accruing lesions and neurodegeneration manifested as the development of brain atrophy, this MRI measure serves as a predictor for the progression of disability in MS. The elevated T2-LV in the postpartum phase showed a trend toward an increased susceptibility to future worsening disability and correlated with a heightened risk of new relapse activity. In light of this, preventing long-term disease progression hinges on the recognition and surveillance of women at the greatest risk of disease activity during the peripartum period, along with consistent therapy with DMTs during pregnancy or the immediate resumption of DMT after delivery. It holds particular relevance for women receiving DMTs, such as fingolimod or Natalizumab, as their discontinuation can precipitate rebound MS activity (McKay et al. 2015).

2.2.4 Effect of Consecutive Pregnancies

Having multiple pregnancies, as opposed to having none, diminishes the likelihood of a woman developing MS. A study in Australia involving 282 MS patients and 542 controls discovered that the risk of experiencing the initial demyelinating event decreased when the number of pregnancies increased (Ponsonby et al. 2012; Voskuhl and Momtazee 2017). Gravidity, or the number of pregnancies, lowers the risk of

MS by 36% (Benoit et al. 2016; Ghajarzadeh et al. 2022). In addition, longer-term investigations tracking women with MS for several years indicated that those who experienced multiple gestations took more time to reach a specific disability level and exhibited less disability (Voskuhl and Momtazee 2017). This is likely related to women with a less disabling MS condition being more inclined to have multiple pregnancies. Another hypothesis suggested is that the rate of disability accumulation might be influenced by recurring exposure to neuroprotective pregnancy hormones at elevated levels during the final trimester (Voskuhl and Momtazee 2017).

A study revealed no correlation between the duration to reach disability and the number of pregnancies. However, it identified that the conversion rate from RRMS, characterized by acute worsening followed by recovery, to secondary-progressive MS (SPMS), marked by steady clinical decline independent of relapses, was higher in women with only one pregnancy (mono-parous) compared to those with multiple pregnancies (multiparous). In a cohort of 973 women from Belgium, multiparity was associated with a reduced risk of progression in RRMS (but not PPMS) compared to nulliparity (D'hooghe et al. 2012). A systematic review by D'Amico et al. (2016) further supported the argument that no link exists between long-term disability and the number of pregnancies. Unlike the systematic review, a negative correlation was, in fact, found between parity and disability score, with birthing mothers having a prolonged time to disability compared to nulliparous women (Alhomoud et al. 2021; Masera et al. 2014).

2.2.5 Long-Term Progression

Pregnancy significantly reduced the risk of long-term MS progression, decreasing it by 50% for each pregnancy. This study (Voskuhl and Momtazee 2017) was especially insightful as it focused on the initial demyelinating event rather than a relapse in a case of established MS. This approach allowed for a clear differentiation between the physiological effect of pregnancy on MS and the possible bias that having MS might exert on a patient's choice to conceive. The increase in the female:male MS incidence from 2 to 3 over the last several decades led to an interesting interpretation of the results. The findings led to the conjecture that the increase in the occurrence of MS in women could be due to the contemporary trend among women to postpone their first pregnancy and have fewer children.

Research involving 102 Iranian women with MS found no significant difference between the ARR during pregnancy and 6 years after delivery. Moreover, no discernible alteration in the mean EDSS was observed in the years following conception (Alhomoud et al. 2021; Keyhanian et al. 2012). Another study focusing on 55 pregnant women with MS found an increase in the average EDSS in the postpartum compared to the pregnancy period, 1.5 and 0.7, respectively. It suggests that, in terms of disability, pregnancy appears to have a protective effect on MS compared to the postpartum period (Alhomoud et al. 2021; Lai et al. 2018).

2.2.6 Obstetric Outcomes

A retrospective study compared 43 deliveries from women with MS to 2975 deliveries in a cohort of women acting as healthy controls. Differences in obstetric outcomes, such as the birth weight, type of delivery, or mean gestational age, in controls versus women with MS were non-substantial. However, some adverse outcomes, including a higher rate of cesarean sections and a decrease in the length and birth weight of newborns from mothers with MS, have been reported (Alhomoud et al. 2021). A Norwegian study indicated that infants born to women with multiple sclerosis tended to have a weight approximately 100 g lower, on average, than infants from other mothers and were more frequently delivered by cesarean section (Holmøy and Torkildsen 2016). A meta-analysis conducted by Finkelsztejn et al. in 2011 reported a prematurity rate of 10% and a risk of low birth weight of 5.8%, findings consistent across all studies. Literature reviews suggest that women with MS do not exhibit a significantly higher risk of obstetric and neonatal complications, apart from a higher prevalence of cesarean sections, prematurity, and low birth weights. The increased postnatal relapse rate appears to be the primary concern for these women (Finkelsztejn et al. 2011).

2.2.7 DMTs and Pregnancy

Managing MS medically during pregnancy and the postpartum period poses challenges due to the potential risks of exposing the fetus in utero and the newborn through breastfeeding to medications (Voskuhl and Momtazee 2017).

Historically, the general advice for pregnant patients with MS was to refrain from using DMTs, as they were not cleared for use. Currently, no definitive guidelines or research studying the effects of DMTs during pregnancy and breastfeeding exists. Yet, some DMTs may exhibit a relatively harmless effect on the developing fetus based on data from current pregnancy exposure registries. This information can be valuable for patients facing decisions, particularly those with a high level of MS disease activity, experiencing challenges in conceiving or preferring not to discontinue DMTs for an extended period.

The sole medication categorized as group B and considered safe for use during pregnancy is Glatiramer acetate (GA). A multicenter, prospective cohort study aimed to assess the effect of GA exposure on the development of particular obstetric events, such as abortion. By tracking a group of 423 pregnant women exposed to GA, it was found that no correlation could be drawn between GA and an increased risk of preterm delivery, spontaneous abortion, or low birth weight (Giannini et al. 2012). However, some studies have suggested a higher incidence of assisted deliveries and congenital defects with GA (Alhomoud et al. 2021).

Interferon-beta (IFN-β) during pregnancy did not result in complications or harm to the fetus. Nonetheless, available clinical data to determine its safety during pregnancy are inconclusive and contradictory. While some studies indicate a potential increase in the abortion rate with IFN-β, the rates did not significantly differ from those observed in the general population. On the contrary, early exposure to IFN-β during pregnancy has been associated with higher rates of miscarriages, assisted vaginal deliveries, and congenital disabilities (Alhomoud et al. 2021). Arguably, the use of GA or IFN-β may be deemed nonessential given the low relapse rates observed during pregnancy, while remaining cognizant of the fact that once IFN-β is resumed, it takes several months to reach its optimal (Alhomoud et al. 2021). On the one hand, the recently updated European prescribing information allows the option to continue IFN-β therapy up to the time of conception, throughout pregnancy, and during breastfeeding. On the other hand, the USA's current prescribing recommendations stipulate that the use of IFN-β during pregnancy should be only if the potential benefit to the mother justifies the potential risk to the fetus (Vukusic et al. 2021).

Predicting a higher relapse rate during pregnancy and the postpartum period is feasible based on pre-pregnancy disease activity, and it is recommended to stabilize with DMTs. The initiation and prescription of MS treatment should take into account family planning and pregnancy counseling (Alhomoud et al. 2021). A summary of DMTs and they pregnancy safety categories can be found in Table 2.1.

2.3 Breastfeeding

2.3.1 Effect of Breastfeeding on MS

The current recommendation from the American Academy of Pediatrics is to breastfeed infants for the initial 6 months exclusively, followed by the incorporation of solid foods and continued breastfeeding for a total of 1 year or more if feasible and desired by the mother (Voskuhl and Momtazee 2017). Until now, it remains to be elucidated whether breastfeeding provides the mother protection from progression in MS activity. However, research has seen reductions in relapse rates (Voskuhl and Momtazee 2017). The mechanism through which breastfeeding protects against relapses remains unclear. Prolactin levels continue to be elevated in breastfeeding individuals, and prolactin has pro-inflammatory properties (Voskuhl and Momtazee 2017). While some studies have suggested that breastfeeding might offer protection against relapse and disability, the topic remains controversial and warrants additional research (Modrego et al. 2021).

Breastfeeding has been shown to confer a protective effect from MS (Holmøy and Torkildsen 2016). A pivotal study on relapse rates and pregnancy demonstrated a significant decrease in relapse rates in the last trimester, but it observed rebound relapses in the period following delivery. It was concluded that breastfeeding had no impact on relapses occurring in the postpartum period. Subsequent analysis revealed that relapse rate before pregnancy was the most reliable predictor of experiencing a postpartum relapse (Voskuhl and Momtazee 2017). For women who had live births, engaging in breastfeeding for a cumulative duration of at least 15 months was linked to a lowered risk of MS/Clinical Isolated Syndrome (CIS) (Langer-Gould et al.

Table 2.1 Clinical trials on disease-modifying therapies (DMT) for the management of MS (*NR* not reported)

Study title	Interventions	Primary outcome measures	Study results
Effects of ocrevus in relapsing multiple sclerosis (NCT04387734)	Ocrelizumab administered according to FDA guidelines and over a period of 1 year vs. other DMTs	Dynamic gait stability and MRI scans	NR
Exploring the immune response to SARS-CoV-2 modRNA vaccines in patients with secondary progressive multiple sclerosis (AMA-VACC)— (NCT04792567)	Siponimod	Achieving seroconversion 1 week after getting the second vaccine	Cellular and humoral responses can still be developed with continuous siponimod therapy; thus, therapy should not be halted for vaccination purposes
Efficacy of fingolimod in de novo patients versus fingolimod in patients previously treated with a first line disease modifying therapy— (NCT01498887)	Assessed fingolimod's efficacy to those who had not previously received DMTs to those who had	Annual relapse rate	ARR was comparable in both groups
A 6-month, randomized, open-label, patient outcomes, safety and tolerability study of fingolimod (FTY720) 0.5 mg/day vs. comparator in patients with relapsing forms of multiple sclerosis— (NCT01216072)	Fingolimod vs. standard MS DMTs	Change from baseline in the global satisfaction subscale of the treatment satisfaction questionnaire for medication	An improvement in the scores from the treatment satisfaction questionnaire for medication when fingolimod was given compared to standard DMTs

(continued)

Table 2.1 (continued)

Study title	Interventions	Primary outcome measures	Study results
Discontinuation of disease modifying therapies (DMTs) in multiple sclerosis (MS)—(NCT03073603)	Discontinuation of DMTs vs. standard of care	Number new MS relapse and/or MRI brain lesion over the study duration	A higher percentage of new relapses and MRI brain lesions were observed in the drug discontinuation group, but the difference was statistically insignificant
A study to evaluate the safety, tolerability, and exploratory efficacy of IMS001 in subjects with multiple sclerosis—(NCT04956744)	Phase 1 study of IMS001, a stem cell therapy, in MS patients with inadequate responses to other treatments, assessing safety and potential to modulate disease course	Safety and tolerability, number of treatment-emergent adverse events	NR
Drug repurposing using metformin for improving the therapeutic outcome in multiple sclerosis patients—(NCT05298670)	MetFORMIN 1000 mg oral tablet and interferon beta-1a	Change in IL17 in both groups as assessed by ELISA	NR
Efficacy and safety of BIIB033 (opicinumab) as an add-on therapy to disease-modifying therapies (DMTs) in relapsing multiple sclerosis (MS)—(NCT03222973)	BIIB033 (opicinumab) vs. placebo	Part 1 of the study uses the overall response score to assess MS patients, combining the expanded disability status scale (EDSS), timed 25-foot walk, and 9-hole peg tests (dominant and nondominant hands) to measure improvement or worsening. Scores range from +4 (improvement) to −4 (worsening) based on set thresholds. Part 2 tracks the number of participants experiencing adverse events and serious adverse events from baseline to week 169	No statistically significant difference in overall response score between groups. Improvements in expanded disability status scale (EDSS), timed 25-foot walk, and 9-hole peg tests were measured but yielded no statistically significant results. Adverse events were comparable in both group, but serious adverse events were slightly high in the placebo group

(continued)

Table 2.1 (continued)

Study title	Interventions	Primary outcome measures	Study results
A post-authorization, long-term study of ozanimod real-world safety—(NCT05605782)	Assessing adverse events in RRMS patients taking Ozanimod to those using other S1P modulators or different non-S1P DMTs	Occurrence of major adverse cardiovascular events, serious opportunistic infection, serious acute liver injury, macular edema, identified rate of malignancies	NR
Impact of disease modifying therapies (DMTs) and associated support services in relapsing multiple sclerosis (RMS) patients—(NCT01601119)	Interferon beta-1a vs. other DMTs	Treatment Satisfaction Questionnaire for Medication-9 questionnaire	NR
Discontinuation of disease modifying therapies (DMTs) in multiple sclerosis (MS): extension of the DISCOMS study—(NCT04754542)	Discontinued medication	New inflammatory disease in group discontinuing DMT vs. those on DMTs, new MRI lesions	NR
Risk perception in multiple sclerosis—(NCT05528666)	High efficacy treatments: Alemtuzumab, ofatumumab, ocrelizumab, natalizumab, cladribine, fingolimod and ozanimod vs. non-high efficacy treatments: Interferons, glatiramer acetate, dimethyl fumarate, and teriflunomide	Proportion of patients who were switched based on risk perception (infections, malignancies, others), proportion of patients who were switched based on risk perception (infections, malignancies, others) were reported. Throughout the study, approximately 5 years (2017–2021)	NR

(continued)

Table 2.1 (continued)

Study title	Interventions	Primary outcome measures	Study results
Tysabri (natalizumab) observational cohort study—multiple sclerosis (MS) registries—(NCT03399981)	Natalizumab vs. other DMTs	This study tracks the number of participants with confirmed progressive multifocal leukoencephalopathy using specific criteria, including brain biopsy or cerebrospinal fluid analysis, and MRI findings. It also records serious adverse events of opportunistic infections in patients switching to Tysabri. The analysis spans retrospective data from before 2016 and prospective data from 2016 to 2023	NR
Impact of the arrival on the French market of new first line oral treatments on the delay between MS onset and first disease modifying treatment (DMTs) administration—(NCT03308994)	Injectables: Interferons and glatiramer acetate vs. oral: Teriflunomide and dimethyl fumarate	MS onset date, first neurological event date. Start of first line DMT date, residence location, and social deprivation score	NR
A study to evaluate long-term safety of Vumerity and Tecfidera in participants with multiple sclerosis (MS)—(NCT05767736)	Assessing incidence of serious adverse events diroximel fumarate, dimethyl fumarate, and other DMTs	Frequency of confirmed serious adverse events	NR
Pregnancy exposure registry for Vumerity (diroximel fumarate)—(NCT05658497)	Assessing major congenital malformations in newborns from mothers taking diroximel fumarate, interferon beta-1a, natalizumab, dimethyl fumarate, or exposed to other DMTs, or unexposed	Number of major congenital malformations	NR

(continued)

Table 2.1 (continued)

Study title	Interventions	Primary outcome measures	Study results
A study to assess pregnancy outcomes in women exposed to diroximel fumarate—(NCT05688436)	Estimates the prevalence of MCMs and other pregnancy outcomes, comparing unexposed to groups to those taking diroximel fumarate, alemtuzumab, fingolimod, glatiramer acetate, interferon beta, natalizumab, ocrelizumab, ofatumumab, ozanimod, peginterferon beta-1a, ponesimod, siponimod	Number of major congenital malformations	NR
Observational study of the effect of ozanimod on fatigue in multiple sclerosis patients—(NCT05319093)	Determine the short-term response to ozanimod by assessing fatigue and brain MRI changes	Change in modified fatigue impact scale score over 3 months	NR

2017). Previous findings in both healthy women and those with MS suggested that extended breastfeeding results in a reduction of pro-inflammatory CD4+ tumor necrosis factor-α (TNF-α)–producing cells. However, cell counts experienced an increase upon the resumption of menstrual cycles (Langer-Gould et al. 2017).

In a prospective study, 201 women with MS were enrolled and followed from pregnancy to 1-year postpartum. The study aimed to investigate the impact of breastfeeding on relapse rate. The breastfeeding women were divided into two groups. Group one consisted of those exclusively breastfeeding with no accompanying or additional feedings for a minimum of 2 months ($n = 120$), while group two consisted of those partially, but not entirely, breastfeeding ($n = 81$). During the initial 6 months after delivery, participants in group one experienced fewer relapses than those in group two (Hellwig et al. 2015; Voskuhl and Momtazee 2017). A reduced number of relapses during pregnancy was noted in participants in group one who had also resumed DMTs in smaller doses in the first 30 days after delivery. This aligns with the hypothesis from the PRIMS study (Vukusic et al. 2004). The hypothesis is that the protective effects of breastfeeding may be influenced by the fact that individuals with increasingly active MS are less likely to breastfeed compared to women with milder MS states (Voskuhl and Momtazee 2017). Highly effective DMTs that are compatible with pregnancy and breastfeeding may be beneficial for women with suboptimal disease control before pregnancy—encouraging women with MS to breastfeed (Langer-Gould et al. 2020).

A randomized clinical trial across multiple centers further supports that breastfeeding leads to lower relapse rates. The study investigated the impact of pregnancy on postpartum relapse rates and found that, compared to women who never breastfed, women with MS who breastfed for more than 3 months experienced a lower incidence of postpartum relapses (Haas and Hommes 2007). Reductions in relapse rates in the initial 6 months postpartum were also observed in another study that focused on women with MS who exclusively breastfed their infants for at least 2 months after delivery (Hellwig et al. 2015). Furthermore, compared to formula feeding, breastfeeding for 6 months or more was linked to a retarded onset of MS (Dalla Costa et al. 2019). These correlations were increasingly prominent among individuals lacking a family history of MS (Alhomoud et al. 2021). Evidently, the duration of breastfeeding and whether an infant receives breast milk or formula can impact the onset of MS relapses.

2.3.2 MS Treatment and Breastfeeding

Most DMTs are not recommended during breastfeeding, with the FDA supporting their discontinuation during this time (Alhomoud et al. 2021). It is noteworthy that this recommendation is more precautionary than scientifically grounded, given the insufficient evidence on infant exposure to the excretion of various MS medications into human breast milk and their negative impact. A common drug used in the treatment of MS is IFN-β1a, which was not found in breast milk. Due to its low plasma levels, substantial protein size, and propensity to bind to its respective receptors, IFN-β1a is less likely to be found in breast milk. Consequently, the continuation of IFN-β1a therapy while breastfeeding could be considered theoretically safe (Voskuhl and Momtazee 2017).

Natalizumab has been detected in breast milk, and its concentration appears to rise by day 50 from the start of treatment, suggesting potential absorption by the infant (Baker et al. 2015; Capone et al. 2022). Consequently, mothers with MS are advised to discontinue breastfeeding while using Natalizumab. Yet, it is still advised that women with MS entirely breastfeed, with an added recommendation that women with active MS before pregnancy or partial recovery from previous relapses should work on optimizing their DMTs prior to conception (Baker et al. 2015; Capone et al. 2022). Women contemplating stopping breastfeeding to restart moderately effective DMTs should be advised that there is no research supporting the reduction of postpartum relapse risk with these DMTs. Additionally, they should be counseled regarding the advantages of breastfeeding on their general health (Langer-Gould et al. 2020).

2.4 Environmental Factors

Research suggests that the development and progression of MS are influenced by a multifaceted interplay between genetic predisposition and distinct environmental factors, collectively elevating susceptibility (Waubant et al. 2019). Broadly, studies have identified potential environmental risk factors that interact with human leukocyte antigen (HLA) susceptibility genes that may contribute to the development or exacerbation of MS. These factors include inadequate sun exposure and low vitamin D levels, obesity, smoking, and infection with the EBV (Waubant et al. 2019).

Research on animal models of MS, or EAE, has aided in our understanding of the immunopathology of MS. Research in mice with EAE found interleukin-23 (IL-23) to be vital in the growth of Th17 cells, which are implicated in the pathogenesis of MS (Jörg et al. 2016). Th17 cells are commonly found in the intestine, where they play a crucial role in immunomodulation by defending against extracellular bacteria and eliminating pathogens. Th17 cells, alongside Th1 cells, are recognized as crucial elements of the CD4+ T effector group that contribute to the pathology in T cell-dependent autoimmune diseases (Jörg et al. 2016). The significance of Th17 cells is also associated with MS, where lesions displayed an increased frequency of interleukin-17 (IL-17)-producing CD4+ T cells. Furthermore, it has been observed that Th17 cells can traverse the inflamed blood–brain barrier (BBB) and release IL-17A, a pro-inflammatory cytokine (Jörg et al. 2016). In contrast to effector T cells, regulatory T cells (Tregs) are central to immunoregulatory responses as they suppress autoreactive immune cells (Jörg et al. 2016). Once activated, Tregs positive for the forkhead box P3 (FoxP3) release anti-inflammatory cytokines such as IL-10 and transforming growth factor (TGF)-β, suppressing autoimmune responses (Jörg et al. 2016). In individuals with MS, an increase in IFN-γ-secreting Tregs was observed. A lower expression of FoxP3 in the Tregs was also observed, along with the inability to suppress inflammation. Hence, it is widely acknowledged that maintaining a balance between Treg subsets and T effector cells plays a significant part in the pathology of autoimmune diseases such as MS (Jörg et al. 2016).

According to Jörg et al. (2016), increasing evidence on the genetic susceptibility of autoimmune conditions explains only part of the risk. This supports the perspective that environmental factors also influence the rising prevalence of MS. We will explore four factors mentioned in epidemiological studies. First, there is clear evidence showing an increased disease incidence in developed countries and higher prevalence with increasing latitude, which will be touched on later in this chapter. Second, reduced sun exposure in high-risk areas, such as Northern Europe, leads to a diminished production of vitamin D, a situation observed in MS patients. Third, ultraviolet (UV) B light increases Treg and tolerogenic dendritic cell levels while decreasing effector T cell levels in patients with MS. Fourth, the hygiene hypothesis posits that individuals that do not come in contact with particular infections from a young age but grow up with effective hygiene may incur an over-functioning immune system, leading to autoimmune disease development. Therefore, infections might confer a protective effect. For example, infections such as EBV increase the risk of MS development when the infection occurs in the later stages of adolescence.

While a comprehensive discussion on the role of vitamin D awaits in Chap. 8, we will briefly touch upon crucial aspects in this section. Vitamin D deficiency and reduced sunlight exposure have long been associated with an elevated risk of MS, prompting investigations into the potential impact of vitamin D supplementation on the disease's progression. Vitamin D acts as an immunomodulator, regulating immune responses and potentially reducing inflammation in MS. It also plays a vital role in bone metabolism, and its synthesis primarily relies on exposure to sunlight's UV radiation. Numerous studies have assessed vitamin D levels in the MS population, revealing correlations between low serum vitamin D levels and an increased risk of MS (Michel 2018; Touil et al. 2023). Some research suggests that vitamin D deficiency in MS patients could stem from issues with the vitamin D receptor (VDR). When activated, the VDR enables vitamin D to initiate a cascade of anti-inflammatory reactions and products (Touil et al. 2023). Vitamin D is also considered neuroprotective, as high doses have been associated with reduced axonal injury (Sangha et al. 2023; Waubant et al. 2019). Adequate vitamin D levels support neuronal health by affecting neurotrophic factor secretion and enhancing neuronal survival (Touil et al. 2023). Vitamin D supplementation and appropriate sun exposure recommendations hold promise in managing MS risk and potentially impacting disease progression. Readers can find more in-depth information on this topic in Chap. 8.

EBV, also known as human herpesvirus 4 (HHV-4), has drawn the attention of MS researchers due to its potential connection to the development of this disease. Research has consistently shown that a significant proportion of individuals with MS have a history of past EBV infection (Pakpoor et al. 2013). More than 90% of people living with MS have encountered an EBV infection at some point in their lives, a higher prevalence than in the general population (Pakpoor et al. 2013). Moreover, there is a 32 times increase in the risk of developing MS in patients with a history of EBV, particularly if the individual experiences symptomatic infectious mononucleosis. Additionally, research has identified specific genetic factors, called HLA-DR2b, with certain variations like HLA-DRB1*1501b and HLA-DRA1*0101a, which can further amplify this risk (Soldan and Lieberman 2023). A comprehensive meta-analysis of Pakpoor et al. (2013) in adult-onset MS revealed that only 1.7% of MS patients tested negative for EBV antibodies, while 6.3% of the control group (those without MS) were EBV seronegative. When data from all 22 studies were synthesized, the overall odds ratio (OR) was calculated to be 0.18 for the risk of developing MS in EBV seronegative adults. This OR indicates a significantly reduced risk of MS in adults without EBV antibodies. Turning our attention to pediatric MS, the study found that 8.3% of pediatric MS patients tested negative for EBV antibodies, whereas a much higher percentage, 34.7%, of controls were EBV seronegative. These findings resulted in an overall odds ratio of 0.18, pointing to a substantially lower risk of MS in children who lacked EBV antibodies (Pakpoor et al. 2013).

2.5 Body Mass Index

Obesity is now acknowledged as a significant risk factor for and contributor to the pathophysiology of MS. Well-conducted observational studies consistently reveal a nearly twofold increased risk of developing pediatric and adult MS when obesity is present during adolescence and early adulthood, compared to individuals with average weight (Alfredsson and Olsson 2019; Waubant et al. 2019). An essential mechanism connecting obesity and MS progression is inflammation. Obesity induces inflammation by raising levels of IL-17 and IL-23, promoting the presence of pro-inflammatory Th17 cells. This chronic inflammation could potentially impact the progression of MS. Additionally, obesity is linked to elevated levels of the inflammatory mediator leptin. Leptin, primarily produced by adipose tissue, influences immune responses and can exacerbate inflammation in obese individuals. Studies have also indicated that increased body weight is associated with decreased levels of vitamin D availability. Vitamin D is crucial in regulating immune responses and possesses anti-inflammatory properties. Its deficiency, often observed in individuals with obesity, can intensify inflammation in the body, potentially worsening the course of MS (Touil et al. 2023).

The involvement of inflammatory pathways and mediators underscores the significance of taking obesity into account during the comprehensive management of patients with MS. In recent years, many MR studies have examined the relationship between adiposity, primarily defined by BMI, and various chronic diseases. MR, characterized by instrumental variable analysis, utilizes genetic variants with a significant impact on the exposure (e.g., BMI) as surrogate markers for the exposure. This approach aims to assess whether the exposure has a causal relationship with the risk of developing a disease. In contrast to traditional observational studies, MR studies are less vulnerable to confounding, as genes are randomly distributed during inheritance. Moreover, MR studies avoid bias from reverse causality since genes remain constant and are not altered by disease development (Larsson and Burgess 2021). Several MR studies have confirmed that BMI levels elevated genetically were associated with an increased risk for MS (Harroud et al. 2021; Mokry et al. 2016; Vandebergh et al. 2022).

The adipose tissue consists of multiple cellular compartments with diverse functions, likely originating from various developmental lineages (Misicka et al. 2023). The body fat is predominantly distributed as visceral adipose tissue (VAT; <20%) or subcutaneous adipose tissue (SAT; 80–90%). SAT is mainly situated under the skin in the femoral (thigh), upper back, abdominal (trunk), and gluteal regions. At the same time, VAT is located close to internal organs within the abdominal cavity. VAT is able to selectively influence metabolism by draining its products through the portal vein. Consequently, VAT and SAT exhibit distinct biological characteristics. For instance, despite possessing different gene expression profiles, both are associated with fasting glucose and triglyceride levels. VAT is associated with increased risk for cardiovascular disease, metabolic syndrome, and mortality, linked to fatty liver, disrupted glucose homeostasis, and systemic oxidative stress. Abdominal SAT has associations with deleterious C-reactive protein, insulin, and lipid levels. In contrast, gluteal-femoral SAT presents a lower risk of metabolic diseases, including type 2 diabetes, cardiovascular diseases, and adverse lipid levels (Misicka et al. 2023). In this MR study, the researchers wanted to understand the relationship between MS risk and measures of obesity, which were SAT, VAT, BMI, waist-hip ratio, and trunk-, leg-, and arm-fat-to-total body fat ratio. BMI is a metric for overall mass, influenced significantly by obesity and excess adipose fat. However, it lacks the ability to distinguish between different tissue types (Misicka et al. 2023).

Nonetheless, the autoimmune activity and systemic inflammation induced in patients suffering from obesity can potentially lead to elevated MS risk. Therefore, a plausible connection between MS and high BMI, mainly due to elevated body fat, particularly VAT, can be drawn. Given that VAT and BMI evaluate distinct yet possibly intersecting aspects of body fat—BMI as a general measure of total body fat and VAT as a specific measure of fat around internal organs—it is plausible that both BMI and VAT may influence the risk of MS via common biological pathways. Several mechanisms can lead to an elevated risk, such as triggering systemic inflammation through the buildup of pro-inflammatory macrophages and B cells and aberrant T cell activation. The study by Misicka et al. (2023) found that only BMI and VAT showed a significant relationship with MS risk, while SAT, waist-hip ratio, and trunk-, leg-, and arm-fat-to-total body fat ratio did not. In conclusion, the heightened risk of MS in individuals with obesity is likely linked to autoimmune activity and systemic inflammation. The study by Misicka et al. (2023) underscores the significant association between MS risk and both BMI and visceral adipose tissue (VAT), suggesting a potential shared biological pathway involving abnormal immune responses and inflammatory processes.

2.6 Smoking

Cigarette smoking is a well-established risk factor associated with the initiation and worsening of MS. The heightened risk is primarily linked to the irritation of lung tissues (Alfredsson and Olsson 2019; Waubant et al. 2019). Some studies suggest that smoking may induce oxidative stress, activate immune cells near the airways, or trigger the aryl hydrocarbon receptor, a known immune response modulator (Waubant et al. 2019). Moreover, smoking leads to post-translational modifications of proteins in the lungs, potentially impacting their antigenicity significantly. The proposed mechanism linking smoking to increased risk of MS suggests a cross-reaction between antigens in the CNS and autoimmune cells and molecules against proteins with post-translational modifications (Hedström et al. 2014). This hypothesis suggests that the risk alteration is due to lung irritation rather than a systemic effect of tobacco. Autoimmune memory cells reside in the lungs, ready to be activated. In EAE studies, these cells show significant proliferation following lung stimulation. Once they develop migratory capabilities, they travel to the CNS and induce inflammation (Hedström et al. 2013). Furthermore, alternative mechanisms propose that smoking directly impacts the CNS by inducing nitric oxide (NO) production, leading to axonal damage and potentially affecting the permeability of the BBB (Correale et al. 2013).

The development of MS is not solely influenced by nicotine commonly found in cigarette smoke. Instead, the development mechanism more likely involves several components present in cigarette smoke. This hypothesis gains support from a study indicating that the risk of MS might be elevated in cases where there is biochemical evidence of passive smoke exposure. An up-and-coming molecular candidate in this context is NO, which is believed to play roles in demyelination and axonal loss. Given cigarette smoke is a well-established risk factor for autoimmune disease development, the likelihood of a biological impact on the immune system is high. Based on animal model studies, it is suggested that exposure to smoke impacts various immune system components, including innate immunity, B- and T-lymphocytes, and natural killer cells (Handel et al. 2011).

The evidence strongly indicates a clear and proportional correlation between smoking and the risk of MS development (Alfredsson and Olsson 2019; Waubant et al. 2019). As an individual's smoking habits increase in both duration and intensity, the corresponding risk of developing MS also rises. Luckily, research has demonstrated that this risk is reversible; if a smoker quits smoking for approximately 10 years, the risk of developing MS can return to a level comparable to that of non-smokers (Michel 2018; Waubant et al. 2019). A critical study conducted in 2001 involving over 100,000 participants in the Nurses' Health Study (NHS) emphasized the correlation between active smoking and a 1.6 times higher incidence of MS compared to non-smokers. This risk increases with cumulative exposure to tobacco. A different cohort study conducted in the UK comprising 443 RRMS and 102 Primary Progressive MS (PPMS) patients investigated the risk of MS onset in individuals who had ever smoked versus those who had never smoked, as per self-reported data. The study found that smoking status had no influence on the susceptibility of developing PPMS compared to RRMS (McKay et al. 2015). Current evidence identifies cigarette smoking as an independent risk factor for MS, with a relative risk of about 1.5. Additionally, smoking is associated with a higher probability of developing progressive disease and experiencing faster disability progression (Wingerchuk 2012).

Additionally, it has been noted that continuous smoking can accelerate the progression of MS, particularly toward SPMS. Moreover, it is associated with more severe disease and a quicker disability progression among individuals with MS (Michel 2018). Intriguingly, smoking elevates the risk of MS regardless of the age of exposure, with both the duration and intensity of smoking independently contributing to this risk (Correale et al. 2013; Michel 2018). Research has also noted that early and more substantial cigarette consumption might enhance the likelihood of developing PPMS compared to onset RRMS characteristics (Mortazavi et al. 2023). Smoking could contribute to a higher rate of conversion from CIS to confirmed MS and accelerate disability accumulation in established progressive forms of MS

(Wingerchuk 2012). The heightened risk of MS associated with current and past smoking is more prominent among individuals with a history of heavy smoking. However, the increased risk of MS linked to smoking diminishes approximately 10 years after smoking cessation, regardless of the timing and cumulative dose of smoking (Hedström et al. 2013).

A distinct dose-response correlation between the cumulative concentration of smoking and the susceptibility of developing MS was identified by research conducted by Hedström et al. (2013). Both the intensity and duration of smoking, rather than the age of smoking onset, appeared to influence the link between smoking and an elevated likelihood of MS development. A twofold heightened possibility of MS development was observed when fewer than five cigarettes were smoked per day over several years. However, the negative impact of smoking diminishes approximately 10 years after quitting smoking, irrespective of the cumulative dose of smoking and age at the time of exposure.

A complex interplay exists between cigarette smoking and genetic factors related to MS risk. Individuals carrying the major MS risk allele HLA-DRB1*15:01, without the protective HLA-A*02 variant, face an increased risk of MS. This risk is further amplified in the presence of smoking, highlighting the intricate interaction between genetic susceptibility and smoking concerning MS (Alfredsson and Olsson 2019; Michel 2018). Epigenetic processes, including DNA methylation, integrate internal and external signals, leading to enduring yet reversible alterations in gene expression. The examination of DNA methylation in blood samples from individuals who smoke has been extensive, revealing numerous loci where methylation levels correlate with smoking intensity and duration since cessation. It has been confirmed that smoking has a genome-wide significant impact on blood DNA methylation in individuals with MS, with a more pronounced effect observed in current smokers and those who ceased smoking within 5 years before sampling. While the effect does not appear to be influenced by the proportions of primary cell types, a noteworthy association with CD34+ regulatory sites was identified. A focused analysis was conducted on the AHRR gene, known for elevated expression in smokers and displaying hypo-methylation at multiple CpG sites (Marabita et al. 2017).

In summary, smoking influences MS regardless of the age when smoking is initiated. A clear association exists between the cumulative dose of smoking and the risk of MS, with both the duration and intensity of smoking independently contributing to the heightened risk. Nevertheless, the adverse impact of smoking gradually diminishes after smoking cessation, irrespective of the total cumulative dose of smoking (Hedström et al. 2013). Collectively, current epidemiological evidence suggests that cigarette smoking, as opposed to other forms of moist tobacco use, is linked to an increased susceptibility to MS. Additionally, a dose-response effect may be present, with a higher consumption of cigarettes corresponding to a greater risk. Significant interactions contributing to the development of MS probably exist between smoking, genetics, and environmental factors (Wingerchuk 2012).

2.7 Geographical Location

The epidemiology of MS exhibits notable variations across regions and populations (Kingwell et al. 2013). Research indicates elevated prevalence (the number of people affected by a disease at or during a particular period) or incidence (the number of new cases within a specified timeframe) figures for MS (Wallin et al. 2019). Over time, an anticipated rise in prevalence estimates can occur with an increase in the life expectancy of individuals with MS. Consequently, the incidence of MS is regarded as a more reliable parameter for tracking changes in MS rates (Kingwell et al. 2013).

MS primarily affects 20–40 years old individuals, with a higher prevalence in females. Globally, there are, on average, 33 individuals with MS per 100,000 people, and this number significantly varies among countries. The highest prevalence is seen in Europe (108/100,000) and North America (140/100,000), while the lowest prevalence is seen in Asia (2.1/100,000) and sub-Saharan Africa (2.2/100,000 people) (Modrego et al. 2021). Differences in the prevalence of MS also exhibit significant variability across different geographical regions. It was suggested that the prevalence of MS increases with higher latitudes, categorizing geographical areas into three regions based on prevalence rates: high (>30/100,000), intermediate (5–25/100,000), and low risk (<5/100,000) (Etemadifar et al. 2013). Historically, Middle Eastern countries were generally labeled as low-risk areas. However, recent evidence in this region classifies it as a higher risk area (Etemadifar et al. 2013). Europe is identified as a high-prevalence region for MS, as defined by Kurtzke's categories of prevalence rates, encompassing over half of the global population diagnosed with MS (Kurtzke 1983; Kingwell et al. 2013). A comprehensive examination of MS epidemiology in the USA revealed a prevalence of 240/100,000 based on a national health survey in Canada (Heydarpour et al. 2015). In Iran, MS prevalence exhibits geographical variation, ranging from 5.3 to 74.28/100,000. Contrary to the geographical distribution hypothesis, research across different provinces suggests Iran is an area of intermediate to high risk (Etemadifar et al. 2013). The prevalence rates reported by Heydarpour et al. (2015) for Middle Eastern and North African countries varied. For example, in 2000, a prevalence of 14.77/100,000 was observed in Kuwait, while Turkey in 2006 observed a prevalence of 101.4/100,000 (Heydarpour et al. 2015). The variations in the burden of MS across regions have been attributed to factors such as socioeconomic status, with regions of higher socioeconomic status having a higher burden of MS. Such findings indicate possible environmental factors such as pollution, sunlight exposure, and smoking as being key players in influencing the degree of MS burden in particular regions (Wallin et al. 2019).

2.8 Conclusion

In summary, the research paper provides insights into the environmental and individual factors influencing MS, particularly highlighting the associations between obesity, smoking, geographical location, and the risk of developing MS. The findings emphasize the complex interplay of genetic, environmental, and lifestyle factors in the development and progression of MS, offering valuable information for further understanding and potentially mitigating the risk of this autoimmune disease.

2.9 Summary

This chapter explores the relationship between MS and pregnancy, examining fertility, the impact of pregnancy on MS, and long-term disease progression. While MS does not significantly reduce fertility, concerns about disease transmission lead to fewer children on average for women with MS. Pregnancy notably reduces MS relapse rates in the third trimester due to hormonal changes, but the postpartum period sees increased relapse risks. This chapter also discusses the effects of multiple pregnancies, showing reduced MS risk but complex long-term progression impacts. Obstetric outcomes for women with MS are generally similar to the broader population, with slight increases in cesarean sections and lower birth weights. DMTs during pregnancy and breastfeeding are reviewed, highlighting the need for careful management due to potential fetal and infant risks. Glatiramer acetate and interferon-beta are noted for their relatively safer profiles. A diverse set of factors such as smoking, vitamin D deficiency, obesity, and Epstein–Barr virus infection are examined for their roles in MS risk and progression. Geographic variations in MS prevalence, linked to sunlight exposure and socioeconomic status, are also discussed. In summary, this chapter provides valuable insights for clinicians and researchers on managing MS during pregnancy, emphasizing the need for tailored family planning and treatment strategies.

References

Alfredsson L, Olsson T (2019) Lifestyle and environmental factors in multiple sclerosis. Cold Spring Harb Perspect Med 9(4):a028944. https://doi.org/10.1101/cshperspect.a028944

Alhomoud MA, Khan AS, Alhomoud I (2021) The potential preventive effect of pregnancy and breastfeeding on multiple sclerosis. Eur Neurol 84(2):71–84. https://doi.org/10.1159/000514432

Baker TE, Cooper SD, Kessler L, Hale TW (2015) Transfer of natalizumab into breast milk in a mother with multiple sclerosis. J Hum Lact 31(2):233–236. https://doi.org/10.1177/0890334414566237

Benoit A, Durand-Dubief F, Amato M-P, Portaccio E, Casey R, Roggerone S, Androdias G, Gignoux L, Ionescu I, Marrosu M-G, Cocco E, Ghezzi A, Annovazzi P, Trojano M, Simone

M, Marignier R, Confavreux C, Vukusic S (2016) History of multiple sclerosis in 2 successive pregnancies. Neurology 87(13):1360–1367. https://doi.org/10.1212/WNL.0000000000003036

Capone F, Albanese A, Quadri G, Di Lazzaro V, Falato E, Cortese A, De Giglio L, Ferraro E (2022) Disease-modifying drugs and breastfeeding in multiple sclerosis: a narrative literature review [review]. Front Neurol 13:851413. https://doi.org/10.3389/fneur.2022.851413

Carbone L, Di Girolamo R, Conforti A, Iorio GG, Simeon V, Landi D, Marfia GA, Lanzillo R, Alviggi C (2023) Ovarian reserve in patients with multiple sclerosis: a systematic review and meta-analysis. Int J Gynecol Obstet 163(1):11–22. https://doi.org/10.1002/ijgo.14757

Confavreux C, Hutchinson M, Hours MM, Cortinovis-Tourniaire P, Moreau T (1998) Rate of pregnancy-related relapse in multiple sclerosis. Pregnancy in Multiple Sclerosis Group. N Engl J Med 339(5):285–291. https://doi.org/10.1056/nejm199807303390501

Correale J, Balbuena Aguirre ME, Farez MF (2013) Sex-specific environmental influences affecting MS development. Clin Immunol 149(2):176–181. https://doi.org/10.1016/j.clim.2013.02.006

D'Amico E, Leone C, Patti F (2016) Offspring number does not influence reaching the disability's milestones in multiple sclerosis: a seven-year follow-up study. Int J Mol Sci 17(2):234

D'hooghe MB, Haentjens P, Nagels G, D'Hooghe T, De Keyser J (2012) Menarche, oral contraceptives, pregnancy and progression of disability in relapsing onset and progressive onset multiple sclerosis. J Neurol 259(5):855–861. https://doi.org/10.1007/s00415-011-6267-7

Dalla Costa G, Romeo M, Esposito F, Sangalli F, Colombo B, Radaelli M, Moiola L, Comi G, Martinelli V (2019) Caesarean section and infant formula feeding are associated with an earlier age of onset of multiple sclerosis. Mult Scler Relat Disord 33:75–77

Etemadifar M, Sajjadi S, Nasr Z, Firoozeei TS, Abtahi SH, Akbari M, Fereidan-Esfahani M (2013) Epidemiology of multiple sclerosis in Iran: a systematic review. Eur Neurol 70(5–6):356–363. https://doi.org/10.1159/000355140

Finkelsztejn A, Brooks J, Paschoal F Jr, Fragoso Y (2011) What can we really tell women with multiple sclerosis regarding pregnancy? A systematic review and meta-analysis of the literature. BJOG Int J Obstet Gynaecol 118(7):790–797. https://doi.org/10.1111/j.1471-0528.2011.02931.x

Ghajarzadeh M, Mohammadi A, Shahraki Z, Sahraian MA, Mohammadifar M (2022) Pregnancy history, oral contraceptive pills consumption (OCPs), and risk of multiple sclerosis: a systematic review and meta-analysis. Int J Prev Med 13:89. https://doi.org/10.4103/ijpvm.IJPVM_299_20

Giannini M, Portaccio E, Ghezzi A (2012) Pregnancy and fetal outcomes after glatiramer acetate exposure in patients with multiple sclerosis: a prospective observational multicentric study, vol 12. BMC Neurol, p 124

Haas J, Hommes OR (2007) A dose comparison study of IVIG in postpartum relapsing-remitting multiple sclerosis. Mult Scler J 13(7):900–908

Handel AE, Williamson AJ, Disanto G, Dobson R, Giovannoni G, Ramagopalan SV (2011) Smoking and multiple sclerosis: an updated meta-analysis. PLoS One 6(1):e16149. https://doi.org/10.1371/journal.pone.0016149

Harroud A, Mitchell RE, Richardson TG, Morris JA, Forgetta V, Davey Smith G, Baranzini SE, Richards JB (2021) Childhood obesity and multiple sclerosis: a Mendelian randomization study. Mult Scler J 27(14):2150–2158. https://doi.org/10.1177/13524585211001781

Hedström AK, Hillert J, Olsson T, Alfredsson L (2013) Smoking and multiple sclerosis susceptibility. Eur J Epidemiol 28(11):867–874. http://www.jstor.org.libconnect.ku.ac.ae/stable/43774910

Hedström AK, Bomfim IL, Barcellos LF, Briggs F, Schaefer C, Kockum I, Olsson T, Alfredsson L (2014) Interaction between passive smoking and two HLA genes with regard to multiple sclerosis risk. Int J Epidemiol 43(6):1791–1798. https://doi.org/10.1093/ije/dyu195

Hellwig K, Rockhoff M, Herbstritt S, Borisow N, Haghikia A, Elias-Hamp B, Menck S, Gold R, Langer-Gould A (2015) Exclusive breastfeeding and the effect on postpartum multiple sclerosis relapses. JAMA Neurol 72(10):1132–1138

Heydarpour P, Khoshkish S, Abtahi S, Moradi-Lakeh M, Sahraian MA (2015) Multiple sclerosis epidemiology in Middle East and North Africa: a systematic review and meta-analysis. Neuroepidemiology 44(4):232–244. https://doi.org/10.1159/000431042

Holmøy T, Torkildsen Ø (2016) Family planning, pregnancy and breastfeeding in multiple sclerosis. Tidsskr Nor Laegeforen 136(20):1726–1729. https://doi.org/10.4045/tidsskr.16.0563. (Familieplanlegging, graviditet og amming ved multippel sklerose)

Hughes SE, Spelman T, Gray OM, Boz C, Trojano M, Lugaresi A, Izquierdo G, Duquette P, Girard M, Grand'Maison F, Grammond P, Oreja-Guevara C, Hupperts R, Bergamaschi R, Giuliani G, Lechner-Scott J, Barnett M, Edite Rio M, van Pesch V et al (2014) Predictors and dynamics of postpartum relapses in women with multiple sclerosis. Mult Scler J 20(6):739–746. https://doi.org/10.1177/1352458513507816

Jörg S, Grohme DA, Erzler M, Binsfeld M, Haghikia A, Müller DN, Linker RA, Kleinewietfeld M (2016) Environmental factors in autoimmune diseases and their role in multiple sclerosis. Cell Mol Life Sci 73(24):4611–4622. https://doi.org/10.1007/s00018-016-2311-1

Keyhanian K, Davoudi V, Etemadifar M, Amin M (2012) Better prognosis of multiple sclerosis in patients who experienced a full-term pregnancy. Eur Neurol 68(3):150–155

Kingwell E, Marriott JJ, Jetté N, Pringsheim T, Makhani N, Morrow SA, Fisk JD, Evans C, Béland SG, Kulaga S, Dykeman J, Wolfson C, Koch MW, Marrie RA (2013) Incidence and prevalence of multiple sclerosis in Europe: a systematic review. BMC Neurol 13:128. https://doi.org/10.1186/1471-2377-13-128

Kurtzke JF (1983) Rating neurologic impairment in multiple sclerosis. Neurology 33(11):1444–1444. https://doi.org/10.1212/WNL.33.11.1444

Lai W, Kinoshita M, Peng A, Li W, Qiu X, Zhu X, He S, Zhang L, Chen L (2018) Does pregnancy affect women with multiple sclerosis? A prospective study in Western China. J Neuroimmunol 321:24–28

Langer-Gould A, Smith JB, Hellwig K, Gonzales E, Haraszti S, Koebnick C, Xiang A (2017) Breastfeeding, ovulatory years, and risk of multiple sclerosis. Neurology 89(6):563–569. https://doi.org/10.1212/wnl.0000000000004207

Langer-Gould A, Smith JB, Albers KB, Xiang AH, Wu J, Kerezsi EH, McClearnen K, Gonzales EG, Leimpeter AD, Van Den Eeden SK (2020) Pregnancy-related relapses and breastfeeding in a contemporary multiple sclerosis cohort. Neurology 94(18):e1939–e1949. https://doi.org/10.1212/wnl.0000000000009374

Larsson SC, Burgess S (2021) Causal role of high body mass index in multiple chronic diseases: a systematic review and meta-analysis of Mendelian randomization studies. BMC Med 19(1):320. https://doi.org/10.1186/s12916-021-02188-x

Marabita F, Almgren M, Sjöholm LK, Kular L, Liu Y, James T, Kiss NB, Feinberg AP, Olsson T, Kockum I, Alfredsson L, Ekström TJ, Jagodic M (2017) Smoking induces DNA methylation changes in multiple sclerosis patients with exposure-response relationship. Sci Rep 7(1):14589. https://doi.org/10.1038/s41598-017-14788-w

Masera S, Cavalla P, Prosperini L, Mattioda A, Mancinelli CR, Superti G, Chiavazza C, Vercellino M, Pinessi L, Pozzilli C (2014) Parity is associated with a longer time to reach irreversible disability milestones in women with multiple sclerosis. Mult Scler J 21(10):1291–1297. https://doi.org/10.1177/1352458514561907

McKay KA, Kwan V, Duggan T, Tremlett H (2015) Risk factors associated with the onset of relapsing-remitting and primary progressive multiple sclerosis: a systematic review. Biomed Res Int 2015:817238. https://doi.org/10.1155/2015/817238

Michel L (2018) Environmental factors in the development of multiple sclerosis. Rev Neurol (Paris) 174(6):372–377. https://doi.org/10.1016/j.neurol.2018.03.010

Misicka E, Gunzler D, Albert J, Briggs FBS (2023) Characterizing causal relationships of visceral fat and body shape on multiple sclerosis risk [article]. Mult Scler Relat Disord 79:104964. https://doi.org/10.1016/j.msard.2023.104964

Modrego PJ, Urrea MA, de Cerio LD (2021) The effects of pregnancy on relapse rates, disability and peripartum outcomes in women with multiple sclerosis: a systematic review and meta-analysis. J Comp Eff Res 10(3):175–186. https://doi.org/10.2217/cer-2020-0211

Mokry LE, Ross S, Timpson NJ, Sawcer S, Davey Smith G, Richards JB (2016) Obesity and multiple sclerosis: a Mendelian randomization study. PLoS Med 13(6):e1002053. https://doi.org/10.1371/journal.pmed.1002053

Mortazavi SH, Moghadasi AN, Almasi-Hashiani A, Sahraian MA, Goudarzi H, Eskandarieh S (2023) Waterpipe and cigarette smoking and drug and alcohol consumption, and the risk of primary progressive multiple sclerosis: a population-based case-control study. Curr J Neurol 22(2):72

Pakpoor J, Disanto G, Gerber JE, Dobson R, Meier UC, Giovannoni G, Ramagopalan SV (2013) The risk of developing multiple sclerosis in individuals seronegative for Epstein-Barr virus: a meta-analysis. Mult Scler 19(2):162–166. https://doi.org/10.1177/1352458512449682

Ponsonby A-L, Lucas R, Van Der Mei I, Dear K, Valery P, Pender M, Taylor B, Kilpatrick T, Coulthard A, Chapman C (2012) Offspring number, pregnancy, and risk of a first clinical demyelinating event: the AusImmune Study. Neurology 78(12):867–874

Sangha A, Quon M, Pfeffer G, Orton S-M (2023) The role of vitamin D in neuroprotection in multiple sclerosis: an update. Nutrients 15(13):2978

Saraste M, Väisänen S, Alanen A, Airas L (2007) Clinical and immunologic evaluation of women with multiple sclerosis during and after pregnancy. Gend Med 4(1):45–55

Sicotte NL, Liva SM, Klutch R, Pfeiffer P, Bouvier S, Odesa S, Wu TJ, Voskuhl RR (2002) Treatment of multiple sclerosis with the pregnancy hormone estriol. Ann Neurol 52(4):421–428

Soldan SS, Lieberman PM (2023) Epstein-Barr virus and multiple sclerosis. Nat Rev Microbiol 21(1):51–64. https://doi.org/10.1038/s41579-022-00770-5

Stuart M, Bergstrom L (2011) Pregnancy and multiple sclerosis. J Midwifery Womens Health 56(1):41–47. https://doi.org/10.1111/j.1542-2011.2010.00008.x

Touil H, Mounts K, De Jager PL (2023) Differential impact of environmental factors on systemic and localized autoimmunity. Front Immunol 14:1147447. https://doi.org/10.3389/fimmu.2023.1147447

Vandebergh M, Becelaere S, CHARGE Inflammation Working Group, Dubois B, Goris A (2022) Body mass index, interleukin-6 signaling and multiple sclerosis: a Mendelian randomization study [original research]. Front Immunol 13:834644. https://doi.org/10.3389/fimmu.2022.834644

Voskuhl R, Momtazee C (2017) Pregnancy: effect on multiple sclerosis, treatment considerations, and breastfeeding. Neurotherapeutics 14(4):974–984. https://doi.org/10.1007/s13311-017-0562-7

Vukusic S, Hutchinson M, Hours M, Moreau T, Cortinovis-Tourniaire P, Adeleine P, Confavreux C, The Pregnancy in Multiple Sclerosis Group (2004) Pregnancy and multiple sclerosis (the PRIMS study): clinical predictors of post-partum relapse. Brain 127(6):1353–1360. https://doi.org/10.1093/brain/awh152

Vukusic S, Michel L, Leguy S, Lebrun-Frenay C (2021) Pregnancy with multiple sclerosis. Rev Neurol 177(3):180–194. https://doi.org/10.1016/j.neurol.2020.05.005

Wallin MT, Culpepper WJ, Nichols E, Bhutta ZA, Gebrehiwot TT, Hay SI, Khalil IA, Krohn KJ, Liang X, Naghavi M, Mokdad AH, Nixon MR, Reiner RC, Sartorius B, Smith M, Topor-Madry R, Werdecker A, Vos T, Feigin VL, Murray CJL (2019) Global, regional, and national burden of multiple sclerosis 1990–2016: a systematic analysis for the Global Burden of Disease Study 2016. Lancet Neurol 18(3):269–285. https://doi.org/10.1016/S1474-4422(18)30443-5

Waubant E, Lucas R, Mowry E, Graves J, Olsson T, Alfredsson L, Langer-Gould A (2019) Environmental and genetic risk factors for MS: an integrated review. Ann Clin Transl Neurol 6(9):1905–1922. https://doi.org/10.1002/acn3.50862

Wingerchuk DM (2012) Smoking: effects on multiple sclerosis susceptibility and disease progression. Ther Adv Neurol Disord 5(1):13–22. https://doi.org/10.1177/1756285611425694

Chapter 3
The Role of Gut Microbiota in the Pathophysiology of Multiple Sclerosis

Hana Al-Ali, Salsabil Zubedi, Sara Aljoudi, Nadia Rabeh, Zakia Dimassi, and Hamdan Hamdan

Abstract Growing evidence of the intricate connections between health and the gut microbiota has prompted a novel way of interpreting health and disease. The connection between the gut microbiota and neurological conditions such as Multiple Sclerosis (MS) was one of several captivating recent discoveries. Promising associations are emerging as researchers probe further into the complexity of the human microbiome, providing new insights into the causes and development of MS. This chapter offers a hypothesis on how the gut microbiota can be a possible trigger for the development of MS to prompt potential new therapeutic avenues for better care for MS patients.

Keywords Multiple Sclerosis (MS) · Microbiota · Dysbiosis · Inflammation · Lcn-2 · Amyotrophic Lateral Sclerosis (ALS) · Autoimmune encephalomyelitis

Abbreviations

AD	Alzheimer's disease
ALS	Amyotrophic lateral sclerosis
$A\beta$	Amyloid beta

H. Al-Ali · S. Zubedi · S. Aljoudi · N. Rabeh · H. Hamdan (✉)
Department of Biological Sciences, College of Medicine and Health Sciences, Khalifa University, Abu Dhabi, United Arab Emirates
e-mail: hamdan.hamdan@ku.ac.ae

Z. Dimassi
Department of Medical Sciences, College of Medicine and Health Sciences, Khalifa University, Abu Dhabi, United Arab Emirates

© The Author(s), under exclusive license to Springer Nature Singapore Pte Ltd. 2024
H. Hamdan (ed.), *Exploring the Effects of Diet on the Development and Prognosis of Multiple Sclerosis (MS)*, Nutritional Neurosciences, https://doi.org/10.1007/978-981-97-4673-6_3

CDMS	Clinically definite MS
CIS	Clinically isolated syndrome
CSF	Cerebrospinal fluid
DMF	Dimethyl fumarate
EAE	Experimental autoimmune encephalomyelitis
GALT	Gut-associated lymphoid tissue
IBD	Inflammatory bowel disease
Lcn-2	lipocalin 2
MS	Multiple Sclerosis
MYD88	Myeloid differentiation primary response protein 88
NGAL	Neutrophil gelatinase-associated lipocalin
NLRP6	NLR pyrin domain-containing 6
NLRs	NOD-like receptors
PBMCs	Peripheral blood mononuclear cells
PRRs	Pattern recognition receptors
SCFAs	Short-chain fatty acids
SFB	Segmented filamentous bacteria
TLRs	Toll-like receptors
TMAO	Trimethylamine N-oxide
Treg	Regulatory T cells

Learning Objectives
- Explore how gut microbiota can trigger MS and its implications for new therapeutic approaches.
- Learn about the importance of biomarkers like fecal lipocalin 2 (Lcn-2) in diagnosing and understanding gut health and inflammation in MS.
- Investigate how imbalances in gut microbiota contribute to systemic inflammation and the progression of MS.
- Compare Gut Microbiomes of MS Patients and Healthy Controls.
- Evaluate how gut microbiota composition influences responses and side effects of MS treatments like dimethyl fumarate (DMF).

3.1 Introduction

Considerable evidence has emerged pointing out that the human microbiome may have a significant influence on the pathophysiology of MS. Research has shown links between particular gut microbiome profiles and certain inflammatory markers, such as fecal lipocalin 2 (Lcn-2). Lcn-2, functioning as an acute-phase protein, acts as a sensitive biomarker for gut health and inflammation, eventually playing a role in the development of MS (Asaf et al. 2023; Yadav et al. 2022).

The developing area of research connecting intestinal inflammation, gut dysbiosis, and MS emphasizes the importance of identifying accurate biomarkers for understanding and treating the condition. The human gut microbiota undergoes rapid changes, particularly under extreme selective stressors like dietary influences, making it susceptible to substantial alterations that may lead to a pathological state known as dysbiosis (Sonnenburg et al. 2016). Multiple studies have shown that the development and progression of MS might be linked to specific microbial taxa (Berer et al. 2017; Chen et al. 2016; Cox et al. 2021; Jangi et al. 2016; Zhou et al. 2022). These microbiomes multiply because of gut dysbiosis, thus contributing to systemic inflammation and overall hastening the course of the disease. However, it is still unclear whether the disease is a result of microbial alterations or vice versa.

3.2 Gut Dysbiosis and Intestinal Inflammation in MS

Gut dysbiosis, derived from Ancient Greek terms δυς/dys, indicating "non-favorable/difficult," and βίος/bios, meaning "way of living," is characterized by an imbalance in the gut microbiota, potentially increasing the risk of certain diseases (Sikalidis and Maykish 2020). It is hypothesized to play a role in triggering systemic inflammation in various neurological and immunological disorders, such as MS, by aggravating symptoms and possibly accelerating the course of the disease. Given the reciprocal communication between the CNS and gut, forming the gut-brain axis, instances of gut dysbiosis can lead to the development or worsening of neurodegenerative conditions such as amyotrophic lateral sclerosis (ALS), Alzheimer's disease (AD) and MS. For example, an abundance of *Parabacteroides distasonis* and *Ruminococcus torques* were associated with worsened ALS symptoms (Dono et al. 2022). Not only can the composition and diversity of the gut microbiota lead to severe consequences, but the metabolites released by these microorganisms have been implicated in the pathogenesis of several neurodegenerative diseases (Solanki et al. 2023). For instance, trimethylamine N-oxide (TMAO), a bacterial metabolite, was found to aggravate the buildup of Amyloid beta (Aβ) plaques, a hallmark of AD development (Solanki et al. 2023). Additionally, autoimmune conditions such as inflammatory bowel disease (IBD) have been shown to develop and worsen because of dysbiosis (Levy et al. 2017). Immunological dysregulation can trigger autoimmunity, brought on by the dysbiotic state, leading to loss of immune tolerance to self-antigens.

Gut dysbiosis and persistent intestinal inflammation increase intestinal barrier permeability, allowing hazardous chemicals to enter the circulation (Yadav et al. 2022). Usually, the transfer of dangerous bacteria and antigens into the bloodstream is prevented by gut microbiota that lines and protects the intestinal barrier (Levy et al. 2017). If the gut microbiota is disturbed, then the intestinal barrier loses its protective effect. This mechanism, also known as "leaky gut," may trigger an immunological/inflammatory chain reaction with resultant activation of the gut-associated lymphoid tissue (GALT), production of pro-inflammatory cytokines, and subsequent local and systemic inflammation. In other body systems, such as the CNS, gut dysbiosis may worsen existing neurological and autoimmune conditions (Levy et al. 2017). To prevent or treat immunological disorders by restoring a healthy gut microbiome, it is crucial to have a thorough understanding of the microbes and metabolites implicated in the disease's pathophysiology.

Recent studies have established that the immune system's functions are greatly influenced by the gut microbiota (Hrncir 2022). The immune system is regulated by certain bacterial species, such as segmented filamentous bacteria (SFB) and Clostridium clusters IV and XIVa (Levy et al. 2017). The communication between microbes and immune cells is facilitated by pattern recognition receptors (PRRs), such as Toll-like receptors (TLRs) and NOD-like receptors (NLRs) (Levy et al. 2017). Deficiencies in TLRs, specifically the myeloid differentiation primary response protein 88 (MYD88) involved in its signaling, were found to harbor a unique gut microbiome (Levy et al. 2017). Specific TLRs, such as TLR5, are beneficial as they protect against dysbiosis. However, certain deficiencies or aberrations in the crosstalk between microbes and immune cells can alter the quantity of specific bacterial taxa in the gut by reducing beneficial microbes while increasing hazardous ones. For example, TLR5-deficient mice showed increased levels of Enterobacteriaceae near the intestinal epithelium (Levy et al. 2017).

In addition to the quantity and diversity of gut microbes, their metabolites can also impact immunological responses. Examples include secondary bile acids and short-chain fatty acids (SCFAs) that function by encouraging the growth of regulatory T cells (Treg), which aids in maintaining immunological tolerance (Levy et al. 2017). Dysbiosis can also lead to reduced secretion of IL-18, which is responsible for epithelial repair, intestinal inflammation, and protection from infections. The precise metabolites implicated have not been precisely elucidated, but they act to inhibit NLR pyrin domain-containing 6 (NLRP6). NLRP6 originally acts to increase IL-18 secretion and is involved in the release of antimicrobial peptides (Levy et al. 2017). A summary of the microbiota implicated in MS can be found in Table 3.1.

3.2.1 Comparison of Gut Microbiome Composition in MS Patients and Healthy Controls

A groundbreaking randomized-controlled trial by Zhou et al. in 2022 examined the complex relationship between gut microbial populations, MS risk, and disease development. By comparing MS patients to their ethnically matched healthy, unrelated cohabitant controls, feces samples were collected from both groups and were used to examine the microbial composition using 16S rRNA sequencing. Researchers compared the two groups for trends, contrasts, and similarities and looked for particular microbial species that can possibly be connected to increasing the risk of acquiring multiple sclerosis (MS). The results of this study showed microbial diversity and composition between the gut microbiomes of the two groups. Therefore, a higher risk of acquiring MS was linked to certain microbial taxa, leading to the conclusion that vulnerability to MS may be pre-determined by the gut microbiota. Specifically, Firmicutes and Bacteroidetes are two important bacterial phyla that exert an effect on inflammation and immunological dysregulation. These factors are pivotal in the progression and onset of MS, and they are associated with various autoimmune disorders (Zhou et al. 2022). *Akkermansia muciniphila* is a group of bacteria that live in the gastric mucosa and breaks down mucin, conserving the

Table 3.1 Summary of the microbiota changes observed in MS

Gut microorganism	Abundance in MS models	Metabolites	Immune/neurological effects	References
Akkermansia muciniphila	Upregulated	SCFA and oligosaccharides	Promotes differentiation of peripheral blood mononuclear cells into TH1 cells, thereby stimulating an inflammatory T-lymphocyte response	Xue et al. (2023)
Prevotella spp.	Downregulated	Phytoestrogen (isoflavones and lignans)	Increases the frequency of CD4$^+$FoxP3$^+$ regulatory T cells in periphery and gut and decreases the frequency of pro-inflammatory IFN-y and IL17-producing CD4 T cells in the CNS	Mangalam and Murray (2019), Shahi et al. (2019)
Blautia	Upregulated	SCFA	Produces acetate, which can cause insulin release and promote metabolic syndromes, including hyperglyceridemia, fatty liver disease, and insulin resistance. Thus, it might indirectly influence MS pathogenesis	Negrotto et al. (2016), Ordoñez-Rodriguez et al. (2023), Perry et al. (2016), Shahi et al. (2017)
Dorea	Upregulated	SCFA	Induce pro-inflammatory IFN-y. metabolize sialic acids and degrade mucin	Crost et al. (2013), Shahi et al. (2017)
Mycoplana	Upregulated	Not reported	Not reported	Chen et al. 2016)
Pseudomonas aeruginosa	Upregulated	Pyocyanin	Cross-reaction between antibodies targeting pseudomonas and myelin basic proteins found in the myelin sheath	Hughes et al. (2003), Lau et al. (2004), Shahi et al. (2017)

integrity of the intestinal barrier and controlling inflammation. Changes in the levels of gastric *Akkermansia muciniphila* may influence immunological responses and contribute to immune dysfunction associated with MS (Zhou et al. 2022). *Prevotella* are another notable species that can alter the gut microbiome composition. Changes in the number of *Prevotella* spp. Have been noted in multiple autoimmune diseases, including MS (Zhou et al. 2022). Their effect on immune system modulation may have an impact on MS progression and risk, but the mechanism is poorly understood. A study by Larsen (2017) revealed the link between *Prevotella* spp. and several other chronic inflammatory diseases such as inflammatory bowel disease and rheumatoid arthritis. People with specific chronic inflammatory disorders had a higher percentage of *Prevotella* spp. in their gut microbiota compared to healthy individuals. *Prevotella* was linked to host immunological responses; it increases pro-inflammatory cytokines and immune cells, particularly Th17 cells. Prevotella helps in Th17 cell development, which increases the release of Th17-related cytokines such as IL-17. When *Prevotella*-induced Th17 responses occur, symptoms of chronic inflammatory disorders are exacerbated, and illness severity is amplified (Larsen 2017).

The abundance of microbial species of the *Blautia, Dorea, Mycolpana,* and *Pseudomonas genera* increased in patients with MS (Shahi et al. 2017). The role of certain species such as those of the *Mycoplana* genus remains to be determined (Shahi et al. 2017). Particular species exhibit an interdependent relationship where the byproducts of one bacterium benefit the growth of the other. This is especially the case for the *Dorea* and *Blautia* genera, where the gases produced by *Dorea* are utilized by *Blautia* to promote its growth (Shahi et al. 2017). The enhancement of MS lesions has been attributed to the activity of *Pseudomonas aeruginosa* (Hughes et al. 2003). An enzyme within *P. aeruginosa*, γ-carboxymuconolactone decarboxylase, shares a similar amino acid sequence, specifically residues 110–124, to that of myelin basic protein, which is the second most abundant protein in the myelin of the CNS (Boggs 2006). The amino acid residues within this range have been shown to induce EAE in rats. Thus, the antibodies against *P. aeruginosa* can cross-react and attack the myelin within the CNS.

3.3 Fecal Lcn-2 as a Sensitive Biomarker

Lcn-2, also known as neutrophil gelatinase-associated lipocalin (NGAL), is an acute-phase protein. It is increased in the presence of inflammation and tissue injury and contributes to controlling the immune response (Yadav et al. 2022). It is released into the intestinal lumen during inflammation and is eventually expelled in feces. Therefore, fecal Lcn-2 levels can be used to reflect the continuing inflammatory processes in the gut. This makes it a potential biomarker for gut dysbiosis and intestinal inflammation in several inflammatory disorders, including MS.

A study enrolled a cohort of MS patients representing different disease phases and severity levels and compared them to healthy controls. Feces samples were collected and analyzed using metagenomic sequencing and 16S rRNA to describe the makeup of the gut microbiota (Yadav et al. 2022). It was found that fecal Lcn-2 levels were elevated in MS patients, especially during disease relapse.

In a combined in vitro and in vivo study, the involvement of the LCN-2 oncogene in progressive MS which impairs remyelination was investigated. It was demonstrated that patients with progressive MS had significantly higher levels of LCN-2 in their cerebrospinal fluid (CSF) (Al Nimer et al. 2016). LCN-2 was also strongly expressed in areas of the central nervous system where demyelination and inflammation were prevalent (Al Nimer et al. 2016). in vitro studies have shown that Lcn-2 may directly inhibit remyelination by interfering with the differentiation and maturation of oligodendrocytes in charge of remyelination. Additionally, pro-inflammatory cytokines such as IL-1 produced as a response to LCN-2's pro-inflammatory effects, were found to hinder the remyelination process. Natalizumab was administered to progressive MS patients to reduce LCN2 levels, but only a mere 13% decrease was observed (Al Nimer et al. 2016).

The amount of LCN-2 in the CSF of patients with clinically isolated syndrome (CIS), a term that describes the first clinical onset of potential MS and typically applies to young adults with acute or subacute onset and rapid evolution of symptoms (Efendi 2015), and early-stage MS vs. healthy controls was studied by M. Khalil, A. Renner, and their colleagues (Khalil et al. 2016). According to the study, CSF LCN-2 levels were considerably higher in patients with CIS versus healthy controls.

The levels of LCN-2 were even higher in early MS patients compared to CIS patients. Notably, it was observed that the higher the LCN-2 levels in the CIS group, the greater the chance of developing clinically definite MS (CDMS) (Khalil et al. 2016). This shows that LCN-2 may be a valuable biomarker for predicting the course of the disease.

3.4 Influence of Gut Microbiota in Developing Autoimmune Encephalomyelitis in Mice Using MS Patient Feces

The study "Gut Microbiota from Multiple Sclerosis Patients Enables Spontaneous Autoimmune Encephalomyelitis in Mice" highlighted the interaction linking the gut microorganisms and the development of autoimmune encephalomyelitis, a condition similar to MS, in mice (Berer et al. 2017). The main goal was to test whether the development of a spontaneous immunological reaction resembling encephalomyelitis in the recipient mice may be caused by gut dysbiosis.

Healthy mice as recipients were used in the study, meaning they had no prior autoimmune diseases (Berer et al. 2017). Feces samples from MS patients were taken to extract the gut microbiota. Subsequently, these microbial populations were transplanted into the recipient mice's gastrointestinal tracts. This step aimed to imitate the microbiological alterations seen in MS patients. The recipient mice were then observed for early indications of the development of autoimmune encephalomyelitis, such as signs of inflammation in the CNS and neurological impairments. Histological examination of the CNS tissues was done after the mice manifested relevant symptoms or at specified time intervals to assess for changes in the intestinal microbiota's stability (Berer et al. 2017). The next step was to examine and compare the gut microbiota of both the recipient mice and the MS patients by evaluating the diversity and makeup of the microbial communities in the gut using 16S rRNA sequencing (Berer et al. 2017). The immune response factors, such as the presence of particular immune cells and cytokines in both the mice and MS patients, were examined to comprehend the impact of the change in microbiota on the immune system. It was found that giving the mice intestinal microbiota of MS patients precipitated the symptoms of autoimmune encephalomyelitis, suggesting a possible direct link between the development of autoimmune responses and the altering gut microbiota composition derived from MS patients.

Another study by Egle Cekanaviciute, Bryan B. Yoo, and their colleagues examined the role of gut bacteria from MS patients in modulating immune responses and aggravating symptoms in mouse models of autoimmune encephalomyelitis, a standard animal model of MS. The researchers gave germ-free mice with no gut microbiota, the gut bacteria of MS patients, and compared them with healthy controls. Animals with gut bacteria from MS patients exhibited more severe symptoms of experimental autoimmune encephalomyelitis (EAE), compared to the group of healthy control mice. It was concluded that the CNS autoimmune responses in mice can become worse by the gut microbiota of MS patients (Cekanaviciute et al. 2017). The authors of this study were also interested in Determining whether gut bacteria affect immune cells. They compared the immune reactions of healthy controls and mice with gut bacteria from MS patients. They noticed that when Human peripheral blood mononuclear cells (PBMCs) were exposed to MS patients' gut microbiota, they exhibited a stronger immunological reaction, where the process of activation of pro-inflammatory Th1 and Th17 T cells was enhanced. Th1 and Th17 pro-inflammatory cells are usually associated with autoimmune reactions . It is hypothesized that MS patients' gut bacteria may alter human T-cell responses and trigger a pro-inflammatory reaction exacerbating MS symptoms (Cekanaviciute et al. 2017).

3.4.1 Gut Microbiota's Potential Effect on How Patients Respond to DMF Treatment and Its Side Effects

Dimethyl fumarate (DMF), also known as Tecfidera, is a medication with delayed release properties often used to treat multiple sclerosis (MS) and was shown to have a positive effect on the intestinal flora composition, according to researchers from the University of Basel and the University Hospital Basel (Diebold et al. 2022). The researchers explored if the gut microbiota's makeup can exert an effect on how patients respond to DMF as well as the side effects they develop. In some patients, the drug has successfully lowered disease activity and relapses. On the other hand, it was found that gut flora can also influence the frequency of the side effects occurring during pharmacological therapy. Three months after initiating treatment with DMF, proportions of pro-inflammatory bacteria vs good bacteria were examined. The authors noticed a decrease in the proportion of pro-inflammatory types of gut bacteria and an increase in the development of "good" bacteria in patients with MS (Diebold et al. 2022).

The study also compared how the gut microbiota can influence the response to the side effects of DMF treatments such as lymphopenia. The researchers included patients with MS who were receiving DMF medication in the study cohort (Diebold et al. 2022). The likelihood of lymphopenia, a documented adverse effect of DMF therapy, was tested. They collected and recorded the full clinical history, severity of the disease, treatment plan, and any adverse effects. Subjects' feces were sampled before beginning the DMF treatment. They examined these samples to determine the microbial makeup (Diebold et al. 2022). The participants' lymphocyte counts were regularly checked throughout the course of DMF treatment. They collected fecal samples throughout the study and analyzed the microbiota using 16S rRNA sequencing (Diebold et al. 2022). The purpose was to observe patterns in the gut's microbial population that may be related to lymphopenia development after DMF therapy.

It was found that the chance of developing lymphopenia while receiving DMF treatment was remarkably correlated with specific gut microbial composition (Diebold et al. 2022). The presence of *Akkermansia muciniphilia* and the absence of *Prevotella copri* at baseline was associated with an increase in susceptibility to lymphopenia. It is suggested that the gut microbial composition may potentially influence how the body reacts to DMF treatment (Diebold et al. 2022).

Another group of researchers explored the potential impact of gut microbiota composition on how patients respond to delayed-release DMF treatment for MS (Storm-Larsen et al. 2019). They observed MS patients treated with DMF for 12 weeks. They observed that DMF medication was associated with a decrease in the *Bifidobacterium* genus at 2 weeks and then an increase in *Faecalibacterium* at 12 weeks (Storm-Larsen et al. 2019). Interestingly, while some subjects showed remarkable changes in the makeup of their gut microbiota, other patients only had minor ones. Compared to patients with no GI symptoms, a cohort of patients experiencing GI symptoms had a reduced abundance of *Bacteroides* at baseline and *Dialister* after 2 weeks of treatment with DMF (Storm-Larsen et al. 2019). However, the results must be interpreted cautiously, given the small sample size (Storm-Larsen et al. 2019).

3.4.2 Clinical Implications and Future Directions

The results of these experiments may have important implications for the future monitoring and treatment of MS. As a result of the findings of microbial patterns that were linked to MS, new diagnostic and prognostic tools can be invented. Identifying biomarkers specific to gut dysbiosis and intestinal inflammation in MS can help monitor the severity of symptoms and the progression of the disease. The biomarkers will aid in the course of the disease by detecting gut-related problems as early as possible and prompt treatment initiation. Tracking fecal Lcn-2 levels can allow the personalization of MS treatment. However, more research is still needed to fully understand the link between fecal Lcn-2 levels and gut dysbiosis in MS. Conducting longitudinal studies and involving larger sample sizes will be key in determining the sensitivity and specificity of Lcn-2 as a biomarker. The investigation of the mechanism driving the gut-brain axis in MS may also provide insight into possible novel therapeutic strategies that target the gut. The findings of these studies may also open new possibilities for innovative therapeutic techniques focusing on the gut microbiota composition to possibly arrest the development of autoimmune encephalomyelitis and similar diseases. They also emphasize using a holistic approach to the research of immunological disorders such as MS by considering genetic factors, environmental components, and gut microbiota. Furthermore, physicians can exercise precision medicine in treating MS patients by measuring the sensitivity to DMF-induced lymphopenia which was found to depend on the nature of their gut microbiota composition. In addition, it is highlighted that altering the gut microbiota is a potential supplementary technique to reduce drug-related side effects and ameliorate treatment outcomes.

3.5 Conclusion

The confluence of research investigating the connection between the gut microbiota and the brain has produced remarkable revelations that could possibly change how autoimmune diseases like Multiple Sclerosis (MS) are understood and treated. These studies conjointly paint a vivid picture of the dynamic relationship between our internal ecosystems and neurological health, from the discovery of fecal Lcn-2 as a potential sensitive biomarker for gut dysbiosis and inflammation in MS patients to the remarkable discovery of the impact of gut microbiota on autoimmune encephalomyelitis and the complex associations between microbiota composition and treatment responses.

3.6 Summary

This chapter explores the relationship between gut microbiota and MS, proposing that gut dysbiosis may trigger MS development. It highlights how specific gut microbiome profiles and biomarkers like fecal Lcn-2 are linked to MS, contributing to inflammation and disease progression. The chapter discusses the gut-brain axis and how imbalances in gut bacteria can worsen neurodegenerative diseases. Studies comparing gut microbiomes of MS patients and healthy individuals identify microbial taxa associated with MS risk. Additionally, the chapter examines how gut microbiota composition affects responses to treatments like DMF and suggests new diagnostic, prognostic, and therapeutic approaches for better MS management.

References

Al Nimer F, Elliott C, Bergman J, Khademi M, Dring AM, Aeinehband S, Bergenheim T, Romme Christensen J, Sellebjerg F, Svenningsson A, Linington C, Olsson T, Piehl F (2016) Lipocalin-2 is increased in progressive multiple sclerosis and inhibits remyelination. Neurol Neuroimmunol Neuroinflam 3(1):e191. https://doi.org/10.1212/NXI.0000000000000191

Asaf S, Maqsood F, Jalil J, Sarfraz Z, Sarfraz A, Mustafa S, Ojeda IC (2023) Lipocalin 2—not only a biomarker: a study of current literature and systematic findings of ongoing clinical trials. Immunol Res 71(3):287–313. https://doi.org/10.1007/s12026-022-09352-2

Berer K, Gerdes LA, Cekanaviciute E, Jia X, Xiao L, Xia Z, Liu C, Klotz L, Stauffer U, Baranzini SE, Kümpfel T, Hohlfeld R, Krishnamoorthy G, Wekerle H (2017) Gut microbiota from multiple sclerosis patients enables spontaneous autoimmune encephalomyelitis in mice. Proc Natl Acad Sci 114(40):10719–10724. https://doi.org/10.1073/pnas.1711233114

Boggs JM (2006) Myelin basic protein: a multifunctional protein. Cell Mol Life Sci 63(17):1945–1961. https://doi.org/10.1007/s00018-006-6094-7

Cekanaviciute E, Yoo BB, Runia TF, Debelius JW, Singh S, Nelson CA, Kanner R, Bencosme Y, Lee YK, Hauser SL, Crabtree-Hartman E, Sand IK, Gacias M, Zhu Y, Casaccia P, Cree BAC, Knight R, Mazmanian SK, Baranzini SE (2017) Gut bacteria from multiple sclerosis patients modulate human T cells and exacerbate symptoms in mouse models. Proc Natl Acad Sci 114(40):10713–10718. https://doi.org/10.1073/pnas.1711235114

Chen J, Chia N, Kalari KR, Yao JZ, Novotna M, Paz Soldan MM, Luckey DH, Marietta EV, Jeraldo PR, Chen X, Weinshenker BG, Rodriguez M, Kantarci OH, Nelson H, Murray JA, Mangalam AK (2016) Multiple sclerosis patients have a distinct gut microbiota compared to healthy controls. Sci Rep 6(1):28484. https://doi.org/10.1038/srep28484

Cox LM, Maghzi AH, Liu S, Tankou SK, Dhang FH, Willocq V, Song A, Wasén C, Tauhid S, Chu R (2021) Gut microbiome in progressive multiple sclerosis. Ann Neurol 89(6):1195–1211

Crost EH, Tailford LE, Le Gall G, Fons M, Henrissat B, Juge N (2013) Utilisation of mucin Glycans by the human gut symbiont Ruminococcus gnavus is strain-dependent. PLoS One 8(10):e76341. https://doi.org/10.1371/journal.pone.0076341

Diebold M, Meola M, Purushothaman S, Siewert LK, Pössnecker E, Roloff T, Lindberg RLP, Kuhle J, Kappos L, Derfuss T, Egli A, Pröbstel A-K (2022) Gut microbiota composition as a candidate risk factor for dimethyl fumarate-induced lymphopenia in multiple sclerosis. Gut Microbes 14(1):2147055. https://doi.org/10.1080/19490976.2022.2147055

Dono A, Esquenazi Y, Choi HA (2022) Gut microbiome and neurocritically ill patients. J Neurocrit Care 15(1):1–11

Efendi H (2015) Clinically isolated syndromes: clinical characteristics, differential diagnosis, and management. Nöro Psikiyatri Arşivi 52(Suppl 1):S1

Hrncir T (2022) Gut microbiota dysbiosis: triggers, consequences, diagnostic and therapeutic options. Microorganisms 10:578

Hughes LE, Smith PA, Bonell S, Natt RS, Wilson C, Rashid T, Amor S, Thompson EJ, Croker J, Ebringer A (2003) Cross-reactivity between related sequences found in Acinetobacter sp., Pseudomonas aeruginosa, myelin basic protein and myelin oligodendrocyte glycoprotein in multiple sclerosis. J Neuroimmunol 144(1):105–115. https://doi.org/10.1016/S0165-5728(03)00274-1

Jangi S, Gandhi R, Cox LM, Li N, Von Glehn F, Yan R, Patel B, Mazzola MA, Liu S, Glanz BL (2016) Alterations of the human gut microbiome in multiple sclerosis. Nat Commun 7(1):12015

Khalil M, Renner A, Langkammer C, Enzinger C, Ropele S, Stojakovic T, Scharnagl H, Bachmaier G, Pichler A, Archelos J, Fuchs S, Seifert-Held T, Fazekas F (2016) Cerebrospinal fluid lipocalin 2 in patients with clinically isolated syndromes and early multiple sclerosis. Mult Scler J 22(12):1560–1568. https://doi.org/10.1177/1352458515624560

Larsen JM (2017) The immune response to Prevotella bacteria in chronic inflammatory disease. Immunology 151(4):363–374. https://doi.org/10.1111/imm.12760

Lau GW, Hassett DJ, Ran H, Kong F (2004) The role of pyocyanin in Pseudomonas aeruginosa infection. Trends Mol Med 10(12):599–606. https://doi.org/10.1016/j.molmed.2004.10.002

Levy M, Kolodziejczyk AA, Thaiss CA, Elinav E (2017) Dysbiosis and the immune system. Nat Rev Immunol 17(4):219–232. https://doi.org/10.1038/nri.2017.7

Mangalam AK, Murray J (2019) Microbial monotherapy with Prevotella histicola for patients with multiple sclerosis. Expert Rev Neurother 19(1):45–53. https://doi.org/10.1080/14737175.2019.1555473

Negrotto L, Farez MF, Correale J (2016) Immunologic effects of metformin and pioglitazone treatment on metabolic syndrome and multiple sclerosis. JAMA Neurol 73(5):520–528. https://doi.org/10.1001/jamaneurol.2015.4807

Ordoñez-Rodriguez A, Roman P, Rueda-Ruzafa L, Campos-Rios A, Cardona D (2023) Changes in gut microbiota and multiple sclerosis: a systematic review. Int J Environ Res Public Health 20(5):4624. https://www.mdpi.com/1660-4601/20/5/4624

Perry RJ, Peng L, Barry NA, Cline GW, Zhang D, Cardone RL, Petersen KF, Kibbey RG, Goodman AL, Shulman GI (2016) Acetate mediates a microbiome–brain–β-cell axis to promote metabolic syndrome. Nature 534(7606):213–217. https://doi.org/10.1038/nature18309

Shahi SK, Freedman SN, Mangalam AK (2017) Gut microbiome in multiple sclerosis: the players involved and the roles they play. Gut Microbes 8(6):607–615. https://doi.org/10.1080/19490976.2017.1349041

Shahi SK, Freedman SN, Murra AC, Zarei K, Sompallae R, Gibson-Corley KN, Karandikar NJ, Murray JA, Mangalam AK (2019) Prevotella histicola, a human gut commensal, is as potent as COPAXONE® in an animal model of multiple sclerosis [original research]. Front Immunol 10:462. https://doi.org/10.3389/fimmu.2019.00462

Sikalidis AK, Maykish A (2020) The gut microbiome and type 2 diabetes mellitus: discussing a complex relationship. Biomedicines 8(1):8. https://www.mdpi.com/2227-9059/8/1/8

Solanki R, Karande A, Ranganathan P (2023) Emerging role of gut microbiota dysbiosis in neuroinflammation and neurodegeneration. Front Neurol 14:1149618

Sonnenburg ED, Smits SA, Tikhonov M, Higginbottom SK, Wingreen NS, Sonnenburg JL (2016) Diet-induced extinctions in the gut microbiota compound over generations. Nature 529(7585):212–215. https://doi.org/10.1038/nature16504

Storm-Larsen C, Myhr K-M, Farbu E, Midgard R, Nyquist K, Broch L, Berg-Hansen P, Buness A, Holm K, Ueland T, Fallang L-E, Burum-Auensen E, Hov J, Holmøy T (2019) Gut microbiota composition during a 12-week intervention with delayed-release dimethyl fumarate in multiple sclerosis—a pilot trial. Mul Scler J Exp Transl Clin 5(4):2055217319888767. https://doi.org/10.1177/2055217319888767

Xue C, Li G, Gu X, Su Y, Zheng Q, Yuan X, Bao Z, Lu J, Li L (2023) Health and disease: Akkermansia muciniphila, the shining star of the gut Flora. Research 6:0107. https://doi.org/10.34133/research.0107

Yadav SK, Ito N, Mindur JE, Kumar H, Youssef M, Suresh S, Kulkarni R, Rosario Y, Balashov KE, Dhib-Jalbut S, Ito K (2022) Fecal Lcn-2 level is a sensitive biological indicator for gut dysbiosis and intestinal inflammation in multiple sclerosis [original research]. Front Immunol 13:1015372. https://doi.org/10.3389/fimmu.2022.1015372

Zhou X, Baumann R, Gao X, Mendoza M, Singh S, Katz Sand I, Xia Z, Cox LM, Chitnis T, Yoon H, Moles L, Caillier SJ, Santaniello A, Ackermann G, Harroud A, Lincoln R, Gomez R, González Peña A, Digga E, Baranzini SE (2022) Gut microbiome of multiple sclerosis patients and paired household healthy controls reveal associations with disease risk and course. Cell 185(19):3467–3486.e3416. https://doi.org/10.1016/j.cell.2022.08.021

# Chapter 4
Western Diet Impact on Multiple Sclerosis

Hana Al-Ali, Salsabil Zubedi, Nadia Rabeh, Sara Aljoudi, Zakia Dimassi, and Hamdan Hamdan

Abstract A set of extreme stressors, such as diet, can overwhelm the human gut microbiota, resulting in a pathological state called dysbiosis. A Western-style diet (WSD), also called a meat-sweet or standard American diet, is defined as the overconsumption of snacks, refined grains, high-fat dairy products, red meat, sweets, high-sugar beverages, conventionally raised animal products, high-fructose products, and fast food. Multiple studies have shown that a WSD can alter the gut microbiota composition from a healthy to a dysbiotic state.

Keywords Multiple Sclerosis (MS) · Western-style diet (WSD) · Systemic inflammation · Pro-inflammatory cells · Experimental autoimmune encephalomyelitis (EAE) · Sodium · Grains · High dairy products · High-fat diet (HFD) · Fructose · Sugar

Abbreviations

BBB   Blood-brain barrier
CNS   Central nervous system
DCs   Dendritic cells

H. Al-Ali · S. Zubedi · N. Rabeh · S. Aljoudi · H. Hamdan (✉)
Department of Biological Sciences, College of Medicine and Health Sciences, Khalifa University, Abu Dhabi, United Arab Emirates
e-mail: hamdan.hamdan@ku.ac.ae

Z. Dimassi
Department of Medical Sciences, College of Medicine and Health Sciences, Khalifa University, Abu Dhabi, United Arab Emirates

© The Author(s), under exclusive license to Springer Nature Singapore Pte Ltd. 2024
H. Hamdan (ed.), *Exploring the Effects of Diet on the Development and Prognosis of Multiple Sclerosis (MS)*, Nutritional Neurosciences,
https://doi.org/10.1007/978-981-97-4673-6_4

EAE	Experimental autoimmune encephalomyelitis
FDA	Food and drug administration
HFD	High-fat diet
IL	Interleukin
LPS	lipopolysaccharide
MS	Multiple Sclerosis
MSIF	Multiple Sclerosis International Federation
NF-κB	Nuclear factor kappa B
RA	Rheumatoid Arthritis
RG	Refined grains
RRMS	Relapse-remitting MS
Th	T helper cell
TLR	Toll-like receptors
TRIF	Toll/IL-1R domain-containing adaptor-inducing IFN-β
WG	Whole grains
WSD	Western-style diet

Learning Objectives
- Understand the impact of a Western-style diet (WSD) on gut microbiota and its role in promoting dysbiosis.
- Identify the connection between gut dysbiosis, systemic inflammation, and the progression of MS.
- Explore the specific dietary components of the WSD that contribute to gut microbiota alterations and inflammation.
- Examine the geographical prevalence of MS and its association with dietary patterns.
- Discuss potential therapeutic strategies and advancements in diagnosing and managing MS through dietary interventions and technology.

4.1 Introduction

Modern-day diets such as the Western-style diet (WSD) have introduced nutrients and foods not previously consumed in the hominin evolution. This shift in our diets has led to poor eating habits that cause systemic inflammation and increased prevalence of various chronic diseases (Clemente-Suárez et al. 2023). The WSD is enriched with ultra-processed food, accelerating the inflammatory response and ultimately promoting demyelination of the central nervous system (CNS). Deviations in normal gut microbiota or dysbiosis are directly linked to numerous inflammatory and autoimmune diseases, with multiple sclerosis (MS) being only one. This chapter will shed light on the epidemiology, pathophysiology, and therapeutic strategies of MS through the lens of the WSD. A considerable focus on the mechanistic pathways, dietary components and their associated inflammatory markers in MS will be explored.

4.2 MS Pathway

A group of researchers studied the connection between dietary habits and gut microbiota composition. They divided African children into two groups (De Filippo et al. 2010). Group one had a hypocaloric diet (662–992 kcal/day) and was rich in fiber (10 g/day). In contrast, group two was fed an energy-dense WSD. The study found that group two microbiota composition was more dysbiotic than group one. Hence, the above study may signify the vital role of having a WSD lifestyle in altering the gut microbiota. An unbalanced microbiota state can have a deleterious effect on the inflammatory condition of the body. A WSD can trigger an increase in the production of interleukin (IL) 17A. Therefore, the peroxisome proliferator-activated receptor-α expression will be downregulated, decreasing the inhibition of fatty acid oxidation (Shen et al. 2017). Hence, this causes an increase in endotoxin lipopolysaccharide (LPS), which activates toll-like receptor (TLR)-4, leading to the stimulation of T helper cell (Th) 17 differentiation. Subsequently, there will be an increase in phosphorylation of the pro-inflammatory transcription factor Nuclear factor kappa B (NF-κB) and inflammatory mediators will be released. As a result, the gastrointestinal barrier permeability will increase, allowing the passage of LPS, T cells, cytokines, and bacteria, leading to systemic inflammation (Park et al. 2015). Altered permeability of the intestinal barrier can affect the prognosis of patients with MS (Fig. 4.1). A group of experimenters demonstrated that dysbiotic microbiota could influence the development of experimental autoimmune encephalomyelitis (EAE). The study results showed that germ-free mice, meaning mice free of microorganisms, showed delays in EAE development compared to specific pathogen-free mice, which are mice free of disease-causing pathogens. These findings highlight the role of the gut microbiota in triggering EAE (Berer et al. 2011).

A study by Cosorich et al. demonstrated that Th17 pro-inflammatory cells that originate from the intestine had been present in humans. Those cells were identified in tissues sampled from patients with a history of relapse-remitting MS (RRMS) using an esophagogastroduodenoscopy. Th17 cells play a crucial role in the pathogenesis of MS and are linked to IL17A, which was observed in the active plaques of MS lesions (Cosorich et al. 2017). Consequently, this indicates that the inflammatory state in MS might be linked to specific microbiota changes.

The progressive translocation of gut microbiota into circulation can cause the inflammatory molecules to reach the brain and increase blood-brain barrier (BBB) permeability. The bacterial translocation may induce the production of pro-inflammatory mediators in the circulation and chronic production in the CNS. Therefore, it promotes neurodegeneration, allowing the bacteria to enter the brain. They do so by either diffusion or transportation across the BBB, producing

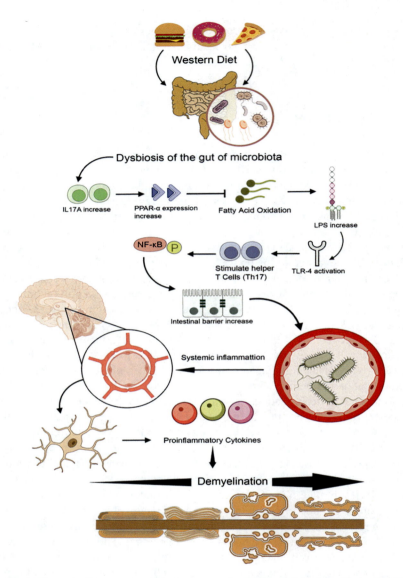

Fig. 4.1 A schematic diagram represents the progression of the pathological effects of WSD on gut microbiota. Several studies reported the critical role of specific inflammatory molecules in deteriorating MS symptoms. First, WSD causes a dysbiotic state in the microbiota, leading to a rise in inflammatory molecules, specifically IL17A, and thus increases the peroxisome proliferator-activated receptor-α expression. Next, there will be a decrease in the inhibition of fatty acid oxidation, raising the level of LPS. Subsequently, TLR-4 will be activated, followed by the stimulation of Th17. The intestinal inflammation leads to a systemic inflammation where the bacteria translocate to the brain, causing neuroinflammation due to increased BBB permeability. Inflammatory cells will enter the brain via diffusion or with the help of a transporter. Therefore, microglial cells in the brain will produce cytokines that will further go on and damage the myelin sheath of neurons (De Filippo et al. 2010; Mirza and Mao-Draayer 2017; Park et al. 2015; Shen et al. 2017). *Th* T helper, *IL* interleukin, *NF-κB* Nuclear factor kappa B, *TLR* Toll-like receptors, *BBB* blood-brain barrier, *LPS* lipopolysaccharide

and activating pro-inflammatory cytokines by microglial cells (Mirza and Mao-Draayer 2017).

In conclusion, having a WSD can worsen the prognosis of patients with MS. This is due to the diet causing dysbiosis of the gut bacteria, increasing intestinal permeability, thus triggering systemic inflammation in the body and increasing BBB permeability. Subsequently, microglial cells become activated, which induces demyelination. However, the gut microbiota is very complex and dynamic, making it challenging to study and give a definite answer.

4.3 Pathway Mechanisms

Diving deeper into the mechanistic pathways involved in MS, although still not fully understood, we can observe similar processes that occur in various inflammatory and autoimmune diseases. The TLR-MyD88 signaling pathway plays a role in activating dendritic cells (DCs), Th17 cells, Th1 cells, and B cells. The TLR-MyD88 and TLR-Toll/IL-1R domain-containing adaptorinducing IFN-β (TRIF) signaling pathways increase the pro-inflammatory cytokine secretion in the body (Zheng et al. 2020). TLR-MyD88 signaling also contributes to the regulation of antigen presentation of DC and affects the integrity of BBB.

The pro-inflammatory IL-17A cytokine released has the destructive effects of causing tissue injuries when overproduced. IL-17A causes inflammation as a response to a disturbance in the body's homeostasis in diseases like MS, Rheumatoid Arthritis (RA), Psoriasis, and others (Khan and Ansar Ahmed 2015).

4.4 High Sodium

The WSD is also known as the high-salt diet because of the high sodium amounts it contains. Several studies showed that large sodium consumption could exacerbate EAE and cause an increase in Th17 cells, leading to increased infiltration of Th17 into the CNS. Other results show that when Th17 cell production increases due to high sodium, it is found to be enriched with pathogenic phenotypes, which modify the adaptive immune system and macrophage response in EAE (Matveeva et al. 2018).

4.5 Refined Grains

Grains are food products that include wheat, oats, rice, cornmeal, cereal grains, and barley products. In WSD, refined grains are ultra-processed products created for longer shelf life. This, however, causes grains to lose their dietary fiber, iron, and many B vitamins. After processing, their texture changes into a finer consistency that lasts longer. They become germ and bran-free post-processing but also lose nutritional value. The lower nutritional quality of refined grains affects the gut microbiota and leads to changes in gut health.

A study conducted by Sally Venegas, Mohsen Meydani, and their colleagues compared the effect of whole grains (WG) and refined grains (RG) on inflammatory response, immune response, gut microbiota, and microbial products in healthy adults. Whole grains were found to have a positive effect on gut microbiota, effector memory T cells, and the acute innate response. RG, on the other hand, resulted in an increase in the pro-inflammatory enterobacteriaceae population, lower stool weight and frequency as compared to WG, and decreased stool output (Vanegas et al. 2017). The findings of this study showcase the direct association between consuming RG and having poor gut health and inflammation.

4.6 High Dairy Products

High dairy products are one of the main components of WSD. They include milk, ice cream, cream, cheeses, and butter. A cross-over study by Casper Swarte, Coby Ealderink, and their colleagues examined the effects of a high-dairy diet vs. a low-dairy diet on gut microbiome in healthy adult subjects. Results show that changes in diet, in short and long-term periods, lead to substantial changes in gut microbiome populations. There was an increase in *S. thermophilus* populations and a reduction in specific butyrate-producing bacteria (*F. prausnitzii*) in the high-dairy diet. Several subjects in this study developed constipation from the high-dairy diet compared to the low-dairy diet (Swarte et al. 2020), further enforcing the relationship between high-dairy diets and dysbiosis.

4.7 High-Fat Diet

A high-fat diet (HFD) is one that consists of 35% of the total calories consumed from either saturated or unsaturated fats. This diet includes processed foods, animal fat, butter, oily fish, and others. High fat in a diet induces obesity and low-grade inflammation due to increased levels of LPS in the circulation, which invokes an immune response. HFD causes dysbiosis, which also leads to obesity and systemic inflammation (Cao et al. 2017). In research studies where animals are genetically prone to obesity, HFDs are used to induce obesity.

4.8 High-Fructose and High-Sugar

High-fructose or high-sugar diets include foods like soft drinks, baked goods, energy drinks, or candy, which have increased significantly over the past 30 years. Consuming large amounts of foods high in sugar or fructose has been proven to stimulate bacteria that produce LPS. When LPS enters blood circulation, an immune response gets triggered, and inflammation occurs (Payne et al. 2012). High fructose and high-sugar diets also lead to an array of health problems such as obesity, increased lipogenesis in the liver, fatty liver disease, systematic inflammation, hypertension, oxidative stress, insulin resistance, and increased adipose tissue mass. A study revealed that high-sugar consumption increases CD4+ cells in EAE mice. A neuroinflammatory response is then triggered due to the disruption of the gut microbiome by high sugar and high fructose (Cao et al. 2017).

4.9 Prevalence

MS prevalence rates follow a certain pattern based on geographical locations. In the past decades, there has been an increase in the prevalence rate in Westernized countries compared to the rest of the world (Matveeva et al. 2018). Researchers have found that the MS incidence rate follows an interesting geographical pattern between and within countries. MS is more common in remote areas, far from the equator, such as North America, Western Europe, and Australasia. MS is the least common in Eastern Saharan Africa and Oceania. Populations farther from the ocean consume diets higher in saturated fats, such as the WSD. Meanwhile, populations who live closer to the ocean and consume more fish tend to be at lower risk of MS (Walton et al. 2020).

The Multiple Sclerosis International Federation (MSIF) has collected data about 115 countries between September 2019 and March 2020. The data represents 87% of the population. Data included adults and children with MS, males and females. Compared to 2013, there was a 30% increase in MS cases in 2020, reaching approximately 2.8 million incidents. Based on the data collected, we observed a prevalence pattern where Europe and the Americas reported the highest incidence rate compared to Southeast Asia and Africa (Walton et al. 2020). These observations indicate that diet does play a role in MS prevalence rates.

4.10 Clinical Implications and Future Directions

Multiple Sclerosis is treated based on symptoms and the disease progression. However, there is a shortage of effective Food and Drug Administration (FDA)-approved symptomatic treatments. MS symptoms are numerous, including fatigue, cognitive disabilities (thinking, learning, and memory), visual problems, muscle spasms/stiffness, neuropathic pain, musculoskeletal pain, emotional problems, sexual problems, bladder problems, bowel problems, speech and swallowing difficulties (Faissner and Gold 2019; Kamm et al. 2014). There are various potential approaches to treating MS symptoms that are being investigated such as targeting the inflammatory response in the body, repairing damaged myelin sheaths, vaccinations, Immune-based therapy, genetic engineering, and others (Johnston Jr and Joy 2001; Ma et al. 2023; Mansilla et al. 2021).

With a disease like MS, however, the earlier the diagnosis, the better the chances of survival and patient outcome. Therefore, efforts are best focused on the development of imaging tools that utilize 3-dimensional conformation, making it easier to evaluate physiological alterations within lesions and surrounding tissues (Bakshi et al. 2008; Kuchling and Paul 2020). This new advanced technology would also help with understanding the disease progression in patients while considering the patient's history. Another positive side of utilizing technology is the ability to visualize information for patients, to help them understand their condition better and focus on improving their lifestyle (Faissner and Gold 2019; Kamm et al. 2014). Focusing on technology and the enhancement of medical resources such as neuroimaging could potentially decrease the time it takes to diagnose patients, which enables physicians to treat symptoms sooner (Banwell et al. 2023).

4.11 Conclusion

Diet is one of the main confounding variables that affect the risk and prognosis of MS. The WSD, heavy with highly processed, nutrient-deficient foods has caused an increase in the prevalence rate of MS in Western countries. Numerous studies show the direct relation between WSD and inflammatory and immune diseases. WSD foods such as refined grains, snacks, high-fat dairy products, red meat, sweets, and fast food have all been studied and linked to MS. They begin by causing a disturbance in gut microbiota, leading to a dysbiotic state, eventually leading to a triggered immune response which attacks itself. There is no single cure for treating MS; however, there are many promising research advances such as utilizing 3D imaging tools and enhancing medical tools to catch and treat early symptoms to slow down disease progression and cause a drastic improvement in the quality of life for patients with MS.

4.12 Summary

This chapter investigates the relationship between a Western-style diet (WSD) and MS, highlighting how the overconsumption of processed foods, high-fat dairy, red meat, and sugars can lead to gut dysbiosis. Dysbiosis triggers systemic inflammation and increases the permeability of the blood-brain barrier, which exacerbates MS symptoms. Studies show that specific dietary components of the WSD promote inflammatory responses, affecting the gut-brain axis and contributing to neurodegeneration. The chapter also discusses the global prevalence of MS, noting higher rates in Westernized countries with prevalent WSD consumption. Lastly, it explores potential therapeutic strategies, including dietary interventions and advanced imaging technologies, to improve MS diagnosis and management.

References

Bakshi R, Thompson AJ, Rocca MA, Pelletier D, Dousset V, Barkhof F, Filippi M (2008) MRI in multiple sclerosis: current status and future prospects. Lancet Neurol 7(7):615–625

Banwell B, Bennett JL, Marignier R, Kim HJ, Brilot F, Flanagan EP, Palace J (2023) Diagnosis of myelin oligodendrocyte glycoprotein antibody-associated disease: international MOGAD panel proposed criteria. Lancet Neurol 22(3):268–282. https://doi.org/10.1016/S1474-4422(22)00431-8

Berer K, Mues M, Koutrolos M, Rasbi ZA, Boziki M, Johner C, Krishnamoorthy G (2011) Commensal microbiota and myelin autoantigen cooperate to trigger autoimmune demyelination. Nature 479(7374):538–541. https://doi.org/10.1038/nature10554

Cao G, Wang Q, Huang W, Tong J, Ye D, He Y, Yin Z (2017) Long-term consumption of caffeine-free high sucrose cola beverages aggravates the pathogenesis of EAE in mice. Cell Discovery 3(1):17020. https://doi.org/10.1038/celldisc.2017.20

Clemente-Suárez VJ, Beltrán-Velasco AI, Redondo-Flórez L, Martín-Rodríguez A, Tornero-Aguilera JF (2023) Global impacts of Western diet and its effects on metabolism and health: a narrative review. Nutrients 15(12):2749

Cosorich I, Dalla-Costa G, Sorini C, Ferrarese R, Messina MJ, Dolpady J, Falcone M (2017) High frequency of intestinal TH17 cells correlates with microbiota alterations and disease activity in multiple sclerosis. Sci Adv 3(7):e1700492. https://doi.org/10.1126/sciadv.1700492

De Filippo C, Cavalieri D, Di Paola M, Ramazzotti M, Poullet JB, Massart S, Lionetti P (2010) Impact of diet in shaping gut microbiota revealed by a comparative study in children from Europe and rural Africa. Proc Natl Acad Sci 107(33):14691–14696. https://doi.org/10.1073/pnas.1005963107

Faissner S, Gold R (2019) Progressive multiple sclerosis: latest therapeutic developments and future directions. Ther Adv Neurol Disord 12:1756286419878323

Johnston RB Jr, Joy JE (2001) Multiple sclerosis: current status and strategies for the future. National Academies Press, Washington, DC

Kamm CP, Uitdehaag BM, Polman CH (2014) Multiple sclerosis: current knowledge and future outlook. Eur Neurol 72(3–4):132–141

Khan D, Ansar Ahmed S (2015) Regulation of IL-17 in autoimmune diseases by transcriptional factors and microRNAs [mini review]. Front Genet 6:236. https://doi.org/10.3389/fgene.2015.00236

Kuchling J, Paul F (2020) Visualizing the central nervous system: imaging tools for multiple sclerosis and neuromyelitis optica spectrum disorders. Front Neurol 11:450

Ma X, Ma R, Zhang M, Qian B, Wang B, Yang W (2023) Recent Progress in multiple sclerosis treatment using immune cells as targets. Pharmaceutics 15(3):728

Mansilla MJ, Presas-Rodríguez S, Teniente-Serra A, González-Larreategui I, Quirant-Sánchez B, Fondelli F, Martínez-Cáceres EM (2021) Paving the way towards an effective treatment for multiple sclerosis: advances in cell therapy. Cell Mol Immunol 18(6):1353–1374. https://doi.org/10.1038/s41423-020-00618-z

Matveeva O, Bogie JFJ, Hendriks JJA, Linker RA, Haghikia A, Kleinewietfeld M (2018) Western lifestyle and immunopathology of multiple sclerosis. Ann N Y Acad Sci 1417(1):71–86. https://doi.org/10.1111/nyas.13583

Mirza A, Mao-Draayer Y (2017) The gut microbiome and microbial translocation in multiple sclerosis. Clin Immunol 183:213–224. https://doi.org/10.1016/j.clim.2017.03.001

Park J-H, Jeong S-Y, Choi A-J, Kim S-J (2015) Lipopolysaccharide directly stimulates Th17 differentiation in vitro modulating phosphorylation of RelB and NF-κB1. Immunol Lett 165(1):10–19. https://doi.org/10.1016/j.imlet.2015.03.003

Payne AN, Chassard C, Lacroix C (2012) Gut microbial adaptation to dietary consumption of fructose, artificial sweeteners and sugar alcohols: implications for host–microbe interactions contributing to obesity. Obes Rev 13(9):799–809. https://doi.org/10.1111/j.1467-789X.2012.01009.x

Shen T, Chen X, Li Y, Tang X, Jiang X, Yu C, Ling W (2017) Interleukin-17A exacerbates high-fat diet-induced hepatic steatosis by inhibiting fatty acid β-oxidation. Biochim Biophys Acta Mol Basis Dis 1863(6):1510–1518

Swarte JC, Eelderink C, Douwes RM, Said MY, Hu S, Post A, Harmsen HJM (2020) Effect of high versus low dairy consumption on the gut microbiome: results of a randomized, cross-over study. Nutrients 12(7):2129

Vanegas SM, Meydani M, Barnett JB, Goldin B, Kane A, Rasmussen H, Meydani SN (2017) Substituting whole grains for refined grains in a 6-wk randomized trial has a modest effect on gut microbiota and immune and inflammatory markers of healthy adults1, 2, 3. Am J Clin Nutr 105(3):635–650. https://doi.org/10.3945/ajcn.116.146928

Walton C, King R, Rechtman L, Kaye W, Leray E, Marrie RA, Baneke P (2020) Rising prevalence of multiple sclerosis worldwide: insights from the atlas of MS. Mult Scler J 26(14):1816–1821. https://doi.org/10.1177/1352458520970841

Zheng C, Chen J, Chu F, Zhu J, Jin T (2020) Inflammatory role of TLR-MyD88 signaling in multiple sclerosis [review]. Front Mol Neurosci 12:314. https://doi.org/10.3389/fnmol.2019.00314

Chapter 5
High Salt Diet Impact on MS

Salsabil Zubedi, Hana Al-Ali, Sara Aljoudi, Nadia Rabeh, Zakia Dimassi, and Hamdan Hamdan

Abstract The influence of sodium intake on autoimmune diseases, particularly Multiple Sclerosis (MS), has garnered significant attention due to its potential role in exacerbating disease activity. Various studies have implicated high-salt diets (HSD) in the pathogenesis of MS, highlighting the link between sodium consumption and immune dysregulation. Mechanistically, elevated salt levels promote the differentiation of pro-inflammatory T helper (Th) 17 cells while compromising the function of regulatory T cells (Tregs), fostering an environment conducive to autoimmunity. Additionally, sodium intake has been associated with the expansion of monocytes and alterations in cytokine profiles, further exacerbating inflammatory responses. However, recent research has provided contrasting findings regarding the direct impact of long-term dietary salt intake on cytokine concentrations in humans. Despite this, animal studies have demonstrated that high sodium intake exacerbates experimental autoimmune encephalomyelitis (EAE), a model for MS, through modulation of gut microbiota and enhancement of Th17 cell differentiation. Notably, the depletion of Lactobacillus species in the gut microbiome due to high salt consumption has been identified as a potential contributing factor to the development of autoimmunity. Understanding the intricate interplay between sodium intake, immune responses, and gut microbiota may pave the way for novel therapeutic strategies to mitigate the detrimental effects of high salt consumption on autoimmune diseases like MS.

Keywords Multiple Sclerosis (MS) · High Salt Diet (HSD) · Sodium Chloride · Salt intake · Inflammation

---

S. Zubedi · H. Al-Ali · S. Aljoudi · N. Rabeh · H. Hamdan (✉)
Department of Biological Sciences, College of Medicine and Health Sciences, Khalifa University, Abu Dhabi, United Arab Emirates
e-mail: hamdan.hamdan@ku.ac.ae

Z. Dimassi
Department of Medical Sciences, College of Medicine and Health Sciences, Khalifa University, Abu Dhabi, United Arab Emirates

© The Author(s), under exclusive license to Springer Nature Singapore Pte Ltd. 2024
H. Hamdan (ed.), *Exploring the Effects of Diet on the Development and Prognosis of Multiple Sclerosis (MS)*, Nutritional Neurosciences,
https://doi.org/10.1007/978-981-97-4673-6_5

Abbreviations

CCR6	C-C chemokine receptor 6
CNS	Central nervous system
EAE	Experimental autoimmune encephalomyelitis
FOXO1	Forkhead box protein O1
GM-CSF	Granulocyte-Macrophage Colony-Stimulating Factor
HSD	High-salt diets
IFN-γ	Interferon-gamma
IL	Interleukin
MPAs	Monocyte platelet aggregates
MS	Multiple Sclerosis
NaCl	Sodium chloride
NFAT5	Nuclear Factor of Activated T cells 5
p38/MAPK	Protein kinase/mitogen-activated protein kinase
RORγt	RAR-related orphan receptor gamma
SGK	Serum Glucocorticoid-Regulated Kinase
Th	T helper
TNFα	Tumor necrosis factor
Tregs	Regulatory T cells

Learning Objectives
- Identify how high-salt diets (HSD) influence autoimmune diseases, particularly multiple sclerosis (MS), and their impact on immune dysregulation.
- Explain the mechanisms by which elevated sodium levels promote pro-inflammatory Th17 cell differentiation and impair regulatory T cell (Treg) function.
- Explore the relationship between high sodium intake, gut microbiota alterations, and the exacerbation of MS-like symptoms in animal models.
- Analyze the contrasting findings regarding the impact of long-term dietary salt intake on cytokine concentrations in humans.
- Discuss potential therapeutic strategies targeting the gut-immune axis to mitigate the detrimental effects of high salt consumption on autoimmune diseases.

5.1 Introduction

Multiple Sclerosis (MS) is a complex autoimmune disease characterized by the immune system's attack on the central nervous system (CNS), leading to demyelination and neurological dysfunction. While genetic factors contribute to susceptibility, emerging research suggests that environmental influences also play a significant role in MS pathogenesis, particularly in genetically predisposed individuals. Among these environmental factors, dietary habits have gained attention, with the Western diet as an illustrative example. The Western diet, characterized by its high sodium content or a pattern of high-salt intake (referred to as a high-salt diet or HSD), has been implicated as a risk factor for various diseases, including MS. Epidemiological evidence suggests that HSD, defined as consumption exceeding 5 g of salt or 2000 mg of sodium per day, may exacerbate MS pathogenesis. The pathological process of MS involves an immune attack against CNS antigens that is mediated by activated CD4 myelin-reactive T cells. HSD has been linked to autoimmune diseases such as MS, and these effects have become increasingly prominent in recent decades (Altowaijri et al. 2017). This trend is likely influenced by shifts in dietary patterns towards higher salt content, particularly in processed and readily available meals.

Recent studies have illuminated the role of pathogenic pro-inflammatory T helper (Th17) cells and pro-inflammatory cytokines in the MS disease process. Research indicates that interleukin (IL)-17-induced CD4+ Th17 cells, particularly of the pathogenic phenotype, play a pivotal role in autoimmune disease development (Zostawa et al. 2017). The authors demonstrated that in vitro introduction of a small quantity of NaCl (40 mM) to a culture of differentiating Th17 cells resulted in a substantial increase in IL-17A in naive CD4 cells. This phenomenon was mediated by p38 mitogen-activated protein kinase/ mitogen-activated protein kinase (p38/MAPK), Nuclear Factor of Activated T cells 5 (NFAT5), and Serum and Glucocorticoid-Regulated Kinase (SGK). Elevated salt consumption induces the differentiation of pro-inflammatory Th17 cells, a process regulated by SGK1 (Katz Sand 2018).

Therefore, increased salt intake promotes the proliferation of pathogenic Th17 cells, exacerbating the course of EAE (Katz Sand 2018). This leads to heightened production of pro-inflammatory cytokines such as Granulocyte-Macrophage Colony-Stimulating Factor (GM-CSF), TNFα, IL-2, interleukin-9, along with various chemokines and C-C chemokine receptor 6 (CCR6) (Zostawa et al. 2017). A study demonstrated that mice exposed to HSD experienced a worsening of EAE, characterized by an elevation in the count of Th17 cells and an increased infiltration of Th17 cells into the CNS (Kleinewietfeld et al. 2013; Zostawa et al. 2017). Research on mice found that elevated sodium levels have been shown to intensify inflammatory responses of both Th2 and potentially Th1 types, compromising the inhibitory capabilities of regulatory T-cells under these conditions (Niiranen et al. 2023). Moreover, the HSD reduces the immunosuppressive effects of Foxp3+ Tregs (regulatory T cells) and heightened interferon-gamma (IFN-γ) secretion by Tregs. Probst and colleagues (Probst et al. 2019) demonstrated similar findings. In vitro, they found that the suppressive function of Tregs was hindered when they were exposed to a high-salt environment as opposed to standard media, while in vivo mice subjected to an HSD exhibited a more severe and rapidly onset disease. Additionally, they demonstrated an elevated percentage of Tregs-producing IFN-γ compared to those on a regular diet.

Similarly, Farez et al. (2015) showed statistically significant higher rates of disease exacerbation in both individuals with moderate (2–4.8 g/day) and high (4.8 g/day or more) sodium intake, at 2.75 and 9.95 times higher, respectively, compared to the low-intake group (under 2 g/day). Moreover, it was illustrated that elevated salt intake augments human serum monocytes, a key player in immunological diseases. It was shown that a brief elevation in salt intake could lead to the expansion of CD14 + CD16- monocytes, along with an upsurge in monocyte platelet aggregates (MPAs) (Zhou et al. 2013). These findings suggest that MPAs could be the cellular foundation for inflammation, leading to risks for thromboembolism and blood pressure fluctuations (Niiranen et al. 2023). In fact, the number of monocytes in the blood of individuals with MS has been significantly higher than in the general population, particularly during disease exacerbations (Probst et al. 2019). The monocytes in individuals with MS tended to transition towards the pro-inflammatory M1 macrophage rather than the anti-inflammatory M2 phenotype, hence inducing an inflammatory state.

5.2 Role of Sodium on Autoimmunity and Inflammation

A recently published study in individuals with MS revealed heightened disease activity and an increased risk of relapse in those with higher dietary sodium intake. The exacerbation of the disease was associated with a greater induction of Th17 cells and an increased presence of pathogenic Th17 cells infiltrating the CNS (Hammer et al. 2017). The authors found that sodium chloride (NaCl) enhanced the development of Th17 cells by roughly 50% compared to the control conditions. Furthermore, NaCl was observed to decrease the differentiation of Th2 cells. Th2 and Th1 cells are a subset of CD4+ T cells that play a crucial role in the immune system. While Th2 cells orchestrate immune responses, Th1 cells respond to intracellular pathogens, such as viruses and certain bacteria. They are involved in cell-mediated immunity, coordinating immune responses that activate cytotoxic T cells and macrophages. An imbalance in the Th1/Th2 ratio can contribute to various immune-related disorders, including autoimmune diseases like MS (Niiranen et al. 2023). Additionally, NaCl was found to enhance the production of the signature pro-inflammatory cytokines IFNγ and IL-17, produced by Th1 and Th17, respectively (Hammer et al. 2017).

5.3 Sodium Intake in MS Activity and Progression

In a study conducted by Niiranen et al. (2023), which was both double-blinded and placebo-controlled, 106 participants were randomly assigned to either the Habitual diet or the Healthy Nordic diet (enriched with berries, vegetables, and fish). Participants were further divided into groups based on either Usual Sodium or Reduced Sodium intake. The findings indicated that adopting the Healthy Nordic Diet enriched with fish and vegetables or reducing salt intake does not alter circulating cytokine concentrations in human subjects.

Cytokine measurements revealed no statistically significant impact on circulating cytokine concentrations between the Reduced Sodium Diet and the Usual Sodium Diet. Remarkably, the Usual Sodium Diet did not elevate levels of the measured signature cytokines associated with Th1 (IFN-γ, IL-12), Th2 (IL-4, IL-13), or Th17 (IL-17, IL-23) lymphocytes. The intriguing findings regarding the salt-induced expansion of Th2 and Th17 cells and their connection to heightened autoimmune reactivity in mice do not appear to have a direct translation to humans in physiologically relevant in-vivo conditions. Hence, this study suggests that acute and exceptionally high salt exposure may lead to noticeable alterations in cytokine concentrations. However, physiologically pertinent changes in long-term diets are unlikely to induce a pro-inflammatory cytokine profile in humans (Niiranen et al. 2023).

5.4 Effect of Sodium on MS Pathophysiology

Salt, or NaCl, has been shown to modulate the pro-inflammatory differentiation of human and mouse Th17 cells (Wu et al. 2013). Despite the potential relationship between sodium consumption and the immune system, there is scarce evidence of its adverse role in autoimmune diseases. However, a study on EAE animal models showed that high sodium intake exacerbates EAE through changes in microbiota and enhancement of Th17 cell differentiation. Recent studies have suggested that the different components of daily diet can affect the frequency of effector T cells in the gut (Haase et al. 2019). In addition, previous data indicate that the pathway of molecules related to sodium balance can influence Th17 cell responses (Wu et al. 2013). In mice, a diet rich in sodium leads to reduced blood flow to the brain and results in cognitive decline, an effect attributed to the elevated production of IL-17, the primary cytokine released by Th17 cells (Niiranen et al. 2023). IL-17, which experiences significant upregulation in chronic CNS lesions in individuals with MS, serves as a marker for demyelination and is indicative of the disease. The heightened infiltration of macrophages and T cells into the CNS, especially around areas of multifocal demyelination, is a substantial indicator of disease severity observed in the clinical presentation of human MS and animal models of the disease (Probst et al. 2019).

5.5 High Salt Intake and Cytokine Levels

The proposed pathway below may link HSD to autoimmune disease in vivo and in vitro (Fig. 5.1). HSD induces the expression of SGK1, which then sends downstream signaling to the IL-23 receptor and promotes its expression. In addition, SGK1 deactivates Forkhead box protein O1 (FOXO1), a repressor of IL-23 receptor expression. Conversely, the RAR-related orphan receptor gamma (RORγt) transcription factor is then free from FOXO1-mediated inhibition on the *IL23* receptor gene. RORγt binds to the *IL23* receptor gene and promotes its transcription, allowing for the expression of the IL-23 receptor. This allows the enhanced differentiation and pathogenic effects of Th17 cells to increase the risk of autoimmunity. Following differentiation, IL-23 cytokines bind to the receptor and trigger inflammation.

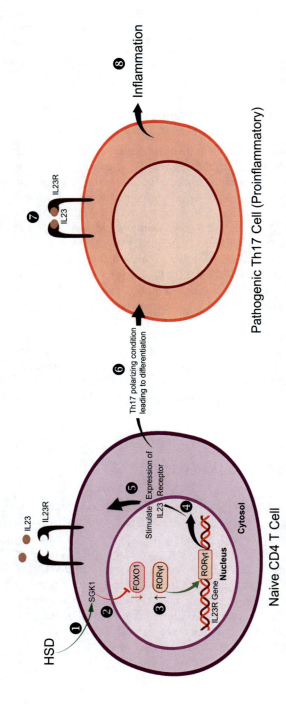

Fig. 5.1 The diagram above illustrates a mechanism linking a high salt diet (HSD) to Th17 pathogenesis, differentiation, and autoimmune disease. A high intake of salt directly affects CD4 T cells. The activation of SGK1 by sodium chloride induces the phosphorylation of FOXO1, a nuclear factor that represses the expression of the *IL23R* gene. Conversely, unopposed RORγt, a transcription factor, is induced due to FOXO1 inhibition, leading to the transcription of the *IL23R* gene. The transcription results in the expression of IL23R at the cell membrane surface, leading to the differentiation of naive CD4 T cells toward the Th17 phenotype. When IL-23 cytokines bind to its receptor, they promote inflammation (Toussirot et al. 2018; Wu et al. 2013).Th T helper, IL23 interleukin 23, IL23R interleukin 23 receptor, HSD high salt diet, SGK1 serum glucocorticoid kinase-1

5.6 Relation of High Salt Content to Gut-Immune Axis

HSD influences the composition of intestinal bacteria in both mice and humans, with a specific suppression observed in *Lactobacillus* spp. Furthermore, HSD affected levels of fecal metabolites in mice, especially those related to bacterial tryptophan metabolites (Wilck et al. 2017). While it is well established that Th17 cell induction is closely dependent on the gut microbiota, the effect of salt on the gut is yet to be fully understood.

In EAE, HSD affects the microbiome by reducing the abundance of *Lactobacillus murinus* (Wilck et al. 2017). One of the functions of this species is the reduction of Th17 cell frequency and expression of IL17A. *L. murinus* is low in 'Western' gut microbiomes, hence the increase in the prevalence of autoimmune diseases in the Western world (Wu et al. 2013). This implies that even modest exposure to salt could impact the continued presence of intestinal Lactobacilli and other bacteria, coupled with an elevation in pro-inflammatory Th17 cells (Wilck et al. 2017). High Salt consumption starting at a young age may give rise to the relative depletion of *Lactobacilli* from the microbiome. Accordingly, this might play a role in the development of autoimmunity. Some studies suggested a moderate salt challenge decreases intestinal *Lactobacilli* and increases pro-inflammatory Th17 cells (Wilck et al. 2017). In conclusion, a diet rich in salt alters the intestinal microbiota and increases intestinal Th17 cells, linking the harmful effects of high salt intake to the gut-immune axis. The recognition of Lactobacillus as a 'natural inhibitor' of Th17 cells induced by high salt intake can be used to develop new treatment strategies.

Intriguingly, newborn infants exhibit the highest levels of Lactobacillus, which decline as they grow older (Wilck et al. 2017). In contrast to the microbiomes of indigenous populations, the prevalence of Lactobacillus in gut microbiomes from 'Western' individuals is relatively scarce. The relationship between HSD and the gut-immune axis can indicate that manipulating gut microbiota could be a prospective strategy to address salt-sensitive conditions.

5.7 Conclusion

The nexus between sodium intake and autoimmune disorders, notably MS, underscores the nuanced interplay among dietary habits, immune responses, and gut microbiota. While existing evidence suggests a potential exacerbating role of HSD in autoimmune reactivity, uncertainties persist regarding the precise mechanisms and long-term implications of sodium consumption on MS progression. Divergent findings regarding the direct impact of dietary salt intake on human cytokine profiles emphasize the imperative for further investigation into the intricate dynamics at play. Nevertheless, animal studies offer compelling insights, linking heightened sodium intake to exacerbated MS-like symptoms through mechanisms implicating gut microbiota modulation and Th17 cell differentiation. Exploring these mechanisms could unveil novel therapeutic strategies aimed at modulating the gut-immune axis to alleviate the adverse effects of excessive salt consumption on autoimmune diseases. As we delve deeper into understanding the complex interplay among sodium intake, immune dysregulation, and gut microbiota, future research endeavors hold promise for elucidating new avenues for managing autoimmune diseases like MS. By leveraging these insights, we can pave the way for innovative interventions and personalized treatments, ultimately improving outcomes for individuals grappling with autoimmune disorders.

5.8 Summary

This chapter examines the influence of high-salt diets on autoimmune diseases, particularly MS. Elevated sodium levels have been shown to promote the differentiation of pro-inflammatory Th17 cells while impairing Tregs, fostering an autoimmune environment. Studies highlight that sodium intake exacerbates MS-like symptoms in animal models by modulating gut microbiota and enhancing Th17 cell differentiation, with a notable depletion of Lactobacillus species in the gut microbiome due to high salt consumption. Although findings regarding the direct impact of long-term dietary salt intake on human cytokine profiles are mixed, animal studies provide compelling evidence linking high sodium intake to increased disease activity. Understanding the interplay between sodium intake, immune responses, and gut microbiota could lead to novel therapeutic strategies to mitigate the adverse effects of excessive salt consumption on autoimmune diseases like MS.

References

Altowaijri G, Fryman A, Yadav V (2017) Dietary interventions and multiple sclerosis. Curr Neurol Neurosci Rep 17(3):28. https://doi.org/10.1007/s11910-017-0732-3

Farez MF, Fiol MP, Gaitán MI, Quintana FJ, Correale J (2015) Sodium intake is associated with increased disease activity in multiple sclerosis. J Neurol Neurosurg Psychiatry 86(1):26–31. https://doi.org/10.1136/jnnp-2014-307928

Haase S, Wilck N, Kleinewietfeld M, Müller DN, Linker RA (2019) Sodium chloride triggers Th17 mediated autoimmunity. J Neuroimmunol 329:9–13. https://doi.org/10.1016/j.jneuroim.2018.06.016

Hammer A, Schliep A, Jörg S, Haghikia A, Gold R, Kleinewietfeld M, Müller DN, Linker RA (2017) Impact of combined sodium chloride and saturated long-chain fatty acid challenge on the differentiation of T helper cells in neuroinflammation. J Neuroinflammation 14(1):184. https://doi.org/10.1186/s12974-017-0954-y

Katz Sand I (2018) The role of diet in multiple sclerosis: mechanistic connections and current evidence. Curr Nutr Rep 7(3):150–160. https://doi.org/10.1007/s13668-018-0236-z

Kleinewietfeld M, Manzel A, Titze J, Kvakan H, Yosef N, Linker RA, Muller DN, Hafler DA (2013) Sodium chloride drives autoimmune disease by the induction of pathogenic TH17 cells. Nature 496(7446):518–522

Niiranen T, Erlund I, Jalkanen S, Jula A, Salmi M (2023) Effects of altered salt intake and diet on cytokines in humans: a 20-week randomized cross-over intervention study. Eur J Immunol 53(1):e2250074. https://doi.org/10.1002/eji.202250074

Probst Y, Mowbray E, Svensen E, Thompson K (2019) A systematic review of the impact of dietary sodium on autoimmunity and inflammation related to multiple sclerosis. Adv Nutr 10(5):902–910. https://doi.org/10.1093/advances/nmz032

Toussirot E, Béreau M, Vauchy C, Saas P (2018) Could sodium chloride be an environmental trigger for immune-mediated diseases? An overview of the experimental and clinical evidence. Front Physiol 9:440. https://doi.org/10.3389/fphys.2018.00440

Wilck N, Matus MG, Kearney SM, Olesen SW, Forslund K, Bartolomaeus H, Haase S, Mähler A, Balogh A, Markó L, Vvedenskaya O, Kleiner FH, Tsvetkov D, Klug L, Costea PI, Sunagawa S, Maier L, Rakova N, Schatz V, Müller DN (2017) Salt-responsive gut commensal modulates TH17 axis and disease. Nature 551(7682):585–589. https://doi.org/10.1038/nature24628

Wu C, Yosef N, Thalhamer T, Zhu C, Xiao S, Kishi Y, Regev A, Kuchroo VK (2013) Induction of pathogenic TH17 cells by inducible salt-sensing kinase SGK1. Nature 496(7446):513–517. https://doi.org/10.1038/nature11984

Zhou X, Zhang L, Ji WJ, Yuan F, Guo ZZ, Pang B, Luo T, Liu X, Zhang WC, Jiang TM, Zhang Z, Li YM (2013) Variation in dietary salt intake induces coordinated dynamics of monocyte subsets and monocyte-platelet aggregates in humans: implications in end organ inflammation. PLoS One 8(4):e60332. https://doi.org/10.1371/journal.pone.0060332

Zostawa J, Adamczyk J, Sowa P, Adamczyk-Sowa M (2017) The influence of sodium on pathophysiology of multiple sclerosis. Neurol Sci 38(3):389–398. https://doi.org/10.1007/s10072-016-2802-8

Chapter 6
From Pasture to Plate: Investigating the Role of Bovine Sources in Multiple Sclerosis

Nadia Rabeh, Sara Aljoudi, Zakia Dimassi, Haya Jasem Al-Ali, Khalood Mohamed Alhosani, and Hamdan Hamdan

Abstract As nutritional science advances, shaping modern dietary guidelines to promote healthy living, the red meat and dairy consumption recommendations continue to spark variability and debate. This paper investigates the nutritional composition and health implications of consuming red meat and cow's milk, particularly in relation to multiple sclerosis (MS). Red meat, abundant in protein, fats, and essential micronutrients, is associated with cardiometabolic diseases and cancer, partly due to the presence of carcinogens formed during processing and cooking. Similarly, cow's milk, while providing essential nutrients, has been linked to autoimmune diseases like MS, potentially exacerbating inflammation and oxidative stress. Preclinical evidence suggests a complex interaction between dietary factors and MS severity, with red meat consumption implicated in iron deposition in MS lesions and the development of autoantibodies. Clinical trials indicate that reducing red meat intake and adopting a Mediterranean diet rich in fruits, vegetables, and fish may alleviate MS symptoms and improve quality of life. However, studies on cow's milk consumption yield mixed results, necessitating further research to elucidate its role in MS risk and severity. Overall, dietary interventions, including reducing red meat consumption and adopting plant-rich diets, hold promise for managing MS symptoms, but additional investigation is needed to develop personalized dietary strategies for individuals with MS, especially considering the potential impact of dietary habits during adolescence on MS risk.

N. Rabeh · S. Aljoudi · H. J. Al-Ali · K. M. Alhosani · H. Hamdan (✉)
Department of Biological Sciences, College of Medicine and Health Sciences, Khalifa University, Abu Dhabi, United Arab Emirates
e-mail: hamdan.hamdan@ku.ac.ae

Z. Dimassi
Department of Medical Sciences, College of Medicine and Health Sciences, Khalifa University, Abu Dhabi, United Arab Emirates

© The Author(s), under exclusive license to Springer Nature Singapore Pte Ltd. 2024
H. Hamdan (ed.), *Exploring the Effects of Diet on the Development and Prognosis of Multiple Sclerosis (MS)*, Nutritional Neurosciences, https://doi.org/10.1007/978-981-97-4673-6_6

Keywords Cow milk · Red meat · Dairy · Multiple sclerosis · Dietary interventions

Abbreviations

BMI	Body mass index
BMMF	Bovine meat and milk factors
BSA	Bovine Serum Albumin
CIDs	Chronic inflammatory diseases
CNS	Central nervous system
CRC	Colorectal cancer
CRP	C-reactive protein
dG	Deoxyguanosine
EAE	Experimental autoimmune encephalomyelitis
EPIC	European Prospective Investigation into Cancer and Nutrition
FCD	First clinical diagnosis of central nervous system demyelination
FFQ	Food Frequency Questionnaire
GGT	Gamma-glutamyltransferase
HAAs	Heterocyclic aromatic amines
hs-CRP	High-sensitivity C-reactive protein
IARC	International Agency for Research on Cancer
IL	Interleukin
LADA	Latent autoimmune diabetes in adults
MAG	Myelin-associated glycoprotein
MOG	Myelin oligodendrocyte glycoprotein
MRI	Magnetic resonance imaging
MR-PheWAS	Phenome-wide Mendelian randomization analysis
MS	Multiple sclerosis
N7-MedG	N7-methyl-2′-deoxyguanosine
NDMA	N-nitrosodimethylamine
NOCs	N-nitroso compounds
O6-MedG	O6-methyl-2′-deoxyguanosine
PAHs	Polycyclic aromatic hydrocarbons
PLCB1	Phospholipase C Beta 1
ppb	Part per billion
PPMS	Primary progressive multiple sclerosis
PUFAs	Polyunsaturated fatty acids
pwMS	Persons with multiple sclerosis
RCTs	Randomized control trials
RRMS	Relapsing-remitting multiple sclerosis
Th	T helper

TNF	Tumor necrosis factor
Treg	T regulatory
US	United States
VLCn3PUFA	Very long-chain omega-3 polyunsaturated fatty acid

Learning Objectives
- Identify the key nutrients in red meat and cow's milk and their potential health implications.
- Explain how red meat and dairy consumption might influence the severity and progression of MS.
- Analyze preclinical studies investigating the relationship between dietary factors, such as red meat and cow's milk, and MS symptoms and severity.
- Assess the findings of clinical trials regarding the impact of reducing red meat intake and adopting plant-rich diets on MS symptoms and quality of life.
- Understand the need for further research to develop personalized dietary strategies for individuals with MS, particularly considering the potential impact of dietary habits during adolescence.

6.1 Introduction

Advancements in scientific understanding of nutrition and its impact on health and disease have helped cultivate today's dietary guidelines to lead healthy lives. These dietary guidelines primarily focus on promoting food groups such as fruits, vegetables, red meat, and dairy products (Herforth et al. 2019). Of these food groups, red meat and dairy recommendations are variable from one guideline to the next. Some guidelines promote their consumption, while others advise limiting it. These variations in dietary recommendations stem from conflicting research regarding red meat and dairy product consumption (Herforth et al. 2019).

Contributing factors to this conflicting evidence include variations in socioeconomic status, culture, and access to these food groups (Godfray et al. 2018). Recent research has pointed to the potential negative health and environmental consequences of excessive red meat consumption (Godfray et al. 2018). Of the adverse health effects currently associated with excessive red meat consumption is colorectal cancer; research is still being pursued to investigate significant associations with other diseases. However, underconsumption of red meat also results in negative health outcomes such as vitamin B12 deficiency (Godfray et al. 2018). On the contrary, evidence has largely supported the consumption of dairy products, with only minimal adverse effects, such as a slightly increased risk of prostate cancer, for which research is inconsistent (Thorning et al. 2016). In light of conflicting evidence, it begs whether diets containing red meat and dairy pose a significant threat to human health and if there is an optimal serving size.

Dietary interventions have recently been studied for the management of Multiple Sclerosis (MS). MS is a neurodegenerative disease where demyelination of the central nervous system (CNS) occurs. In a review of randomized control trials (RCTs) investigating the effect of diet on MS, most studies exclude consuming dairy products. There is also limited evidence solely investigating the effect of red meat intake on MS severity (Parks et al. 2020). The effect of red meat and dairy consumption on MS is currently under study, with current evidence having divided stances on dairy consumption having negative versus positive effects on MS (Ashtari et al. 2013; Chunder et al. 2023; Harirchian et al. 2016; Malosse and Perron 1993; Temperley et al. 2023). Likewise, for red meat, some research has cited improvements in MS with red meat consumption, while others noted increased MS prevalence (Jahromi et al. 2012). In this chapter, the nutritional value of red meat and dairy products, with a focus on cow's milk and their effect on MS, is explored.

6.2 Nutritional Value and Health Implications of Red Meat and Cow's Milk

The term "red meat" is commonly used to refer to groups of meats such as veal, lamb, and beef. Red meat can be consumed fresh or preserved by freezing (Williamson et al. 2005). On the other hand, processed meats refer to meats that have been preserved through methods other than freezing or processing to enhance their flavor profile. For example, processes such as smoking, air-drying, and salting are among some of the methods used to produce processed meats such as bacon, salami, ham, and tinned meat (Farvid et al. 2021; Williamson et al. 2005). With respect to its nutritional value, red meat has an extensive micro-and macro-nutrient contribution (Williamson et al. 2005). Listing precise values for each macro-and micronutrient studied in samples of red meat varies differently among sources. Therefore, a general discussion of the nutritional benefits of red meat will be highlighted.

In addition to being a source of energy, protein in red meats is an important and significant contributor to the growth and repair of the human body. Approximately 20 g of protein is found per 100 g of uncooked red meat, which increases to about 35 g after cooking (Williamson et al. 2005). Protein is composed of amino acids, and different food sources may be rich in particular amino acids or contain essential ones (P. Williams 2007). In the case of red meat, its consumption is recommended as a macronutrient for its ability to provide a set of essential amino acids: isoleucine, threonine, methionine, tryptophan, phenylalanine, leucine, valine, and lysine. These amino acids are essential as they are not endogenously manufactured by the body and can only be obtained from exogenous sources, primarily through a balanced diet (Williams 2007). Of these amino acids, tryptophan has been implicated in immune regulatory processes (Correale 2021). Another important macronutrient red meat provides is fat. Along with helping to make food increasingly palatable, fat is another rich source of energy. It also supplies a vital set of micronutrients, such as the fat-soluble vitamins A, D, E, and K (Williamson et al. 2005). The associated benefits of vitamins A, D, and E are thoroughly discussed in other chapters of this book. Vitamin K plays a key role in maintaining a supply of clotting factors by γ-carboxylation and the protein osteocalcin, which is involved in bone growth (Mladěnka et al. 2022). Fat from red meat is also a source of fatty acids such as polyunsaturated fatty acids (PUFAs), saturated fatty acids, and monounsaturated fatty acids (Williams 2007; Williamson et al. 2005). Depending on the cut and type of red meat, a different amount and type of fatty acids will be seen. The amount and content of fatty acids will also present with positive or negative health outcomes (Williams 2007; Williamson et al. 2005).

The nutritional value of red meat changes based on the source and the mode of processing. Processed meats have been found to be associated with cardiometabolic diseases such as obesity, cardiovascular disease, and type 2 Diabetes Mellitus (Hobbs-Grimmer et al. 2021). Processed red meat is said to contribute to the pathogenesis of these diseases due to their significant high-fat content, as well as the high amounts of salt, which are enriched with nitrosamine compounds for preservation purposes (Hobbs-Grimmer et al. 2021; Pan et al. 2022). Yet, this is not to say that unprocessed red meats are better than processed red meats, as unprocessed meats have also been found to be associated with cardiovascular disease (Micha et al. 2012). The nutritional value of processed meats and unprocessed meats are generally similar, except that the latter have a higher fat and salt content and a higher proportion of nitrates (Micha et al. 2012).

Red meat and processed meat consumption have been linked to increased risk of cancer. The oxidative stress and increased inflammatory markers, such as c-reactive protein (CRP), have been associated with excess red meat consumption and hence elevated cancer risk (Farvid et al. 2021). Additionally, evidence of immunotoxicity and inflammation have been associated in certain parts of the world, such as the United States, where sex hormones like estrogen for cattle growth, have been used, as their metabolites can potentially produce free radicals. In 2015, the International Agency for Research on Cancer (IARC) reported that processed meat is classified as carcinogenic to humans (Group 1), while red meat is classified as probably carcinogenic to humans (Group 2A) (Farvid et al. 2021). The processing and cooking of red meat have been associated with the formation of certain carcinogens such as N-nitroso compounds (NOCs), heterocyclic aromatic amines (HAAs), and polycyclic aromatic hydrocarbons (PAHs). These carcinogens are implicated as possible contributors to colorectal cancer (CRC) (Farvid et al. 2021). These chemicals can form DNA adducts, potentially leading to mutations and cancer development. NOCs include N-nitrosodimethylamine (NDMA), N-nitrosodiethylamine, N-nitrosodibutylamine, N-nitrosopyrrolidine, and N-nitrospiperidine. The levels of NOCs in cured meats vary from less than one part per billion (ppb) to 130 ppb. These compounds undergo metabolic activation by cytochrome P450 2E1 in the gastrointestinal tract (Farvid et al. 2021). For instance, the reactive methylating intermediate of NDMA results in the formation of N7-methyl-2′-deoxyguanosine (N7-MedG), causing abasic site formation, DNA strand breaks, and cytotoxicity. Another adduct is formed through O6-methylation of deoxyguanosine (dG) to produce O6-methyl-2′-deoxyguanosine (O6-MedG) (Farvid et al. 2021). HAAs undergo metabolic processes facilitated by cytochrome P450 enzymes, resulting in the creation of genotoxic N-hydroxylated metabolites, which are further metabolized by conjugation enzymes like N-acetyltransferases or sulfotransferases, ultimately producing reactive intermediates that bind to DNA (Farvid et al. 2021). Similarly, PAHs, such as benzo[a]pyrene, undergo bioactivation through human cytochrome P450 enzymes, leading to the formation of genotoxic species. The reactive intermediates, including anti-diol-epoxides in the bay region of PAH molecules, contribute to DNA damage. Moreover, trans-dihydrodiols of certain PAHs yield o-quinones that can induce DNA damage and oxidative stress (Farvid et al. 2021).

Milk consumption has been a part of the human diet for many years, but recently this trend has decreased. Specifically pertaining to cow's milk, milk is a source of minerals, protein, fats, and sugars. Additional components such as hormones, growth factors, and immunoglobulins can also be found in cow's milk (Haug et al. 2007). Approximately 5% of milk comprises sugar lactose, which is broken down by enzymes into glucose and galactose for absorption in the intestines (Marangoni et al. 2019). A key component in milk is protein, albeit in low amounts. Caseins, the most abundant milk proteins, are rich in the amino acids arginine, valine, leucine, proline, isoleucine, and glutamate (Marangoni et al. 2019). Whey protein is another milk protein with smaller amounts than caseins. Whey protein is also a source of amino acids such as leucine, lysine, tryptophan, and cysteine (Marangoni et al. 2019). In terms of fat composition, lipids comprise approximately 4% of milk, most of which are saturated fatty acids. The most abundant saturated fatty acid is palmitic acid. Unsaturated fatty acids such as oleic, linoleic, and α-linoleic acids are also found in milk. The odd-chain fatty acids are unique as their presence in the body is a marker of milk consumption (Marangoni et al. 2019). In addition to milk's supply of macronutrients, the approximately 1.2% of micronutrients found in milk include calcium, magnesium, phosphorus, zinc, selenium, and potassium. As equally as important as the minerals in milk are the vitamins that it offers. Fat-soluble vitamins such as D, E, and A and water-soluble vitamins B12 and B2 are crucial for optimal health and cellular functioning (Marangoni et al. 2019).

Milk can also be differentiated into pasteurized and unpasteurized forms. Pasteurization is the heat treatment process that aims to decrease pathogenic microbial populations to a level safe for consumption (Melini et al. 2017). The two most important organisms targeted during this process are *Coxiella burnettii* and *Mycobacterium tuberculosis*. Research shows that pasteurizing milk does not alter lactose or fatty acid content (Melini et al. 2017), and only minor changes affect proteins such as caseins and whey, with whey protein undergoing denaturation as a result (Melini et al. 2017). It was found that heating leads to a loss in approximately 1–4% of lysine, while other amino acids were not significantly reduced after pasteurization. Minerals, such as those mentioned above, remain intact after pasteurization. Of all the nutritional components found in milk, vitamin content is significantly impacted upon pasteurization (Melini et al. 2017), with a detectable decrease in vitamins such as B12, B2, C, and E. An increase in vitamin A has been associated with milk pasteurization. However, milk is not a primary source of these vitamins as their concentration before pasteurization is already low. Therefore, the concern regarding the loss of vitamins and their health impacts is trivial (Melini et al. 2017). Nonetheless, milk has been fortified with particular vitamins, such as vitamin D, to mitigate nutritional deficiencies.

Research on the potential health consequences of milk consumption found a link with the progression of autoimmune diseases (Vojdani 2014). There is no single, precise mechanism by which this happens, as it varies for different diseases.

However, it is said that amino acids or peptides in milk mimic host molecules and cross-react with various immune cells to, unfortunately, result in autoimmune disease (Vojdani 2014). Individuals with a milk allergy can enter a pro-inflammatory state upon milk consumption (Rodop et al. 2021). Pro-inflammatory cells such as T helper (Th) 2 increase, while immunomodulating T regulatory (Treg) cells decrease. Consequently, aberrant activation of malfunctioning T and B cells ensues, leading to autoimmunity (Rodop et al. 2021). Despite research demonstrating a link between milk consumption and autoimmunity, clinical trials investigating the effect of dairy consumption on the levels of inflammatory biomarkers did not note a significant association, as there was no documented substantial change in inflammatory markers such as tumor necrosis factor (TNF)-α, interleukin (IL)-6, IL-1, and CRP with milk consumption (Hess et al. 2021). Evidently, additional research is needed to understand the cellular and molecular effects of milk consumption and its implication in autoimmune diseases.

6.3 Preclinical Evidence for the Influence of Dairy and Red Meat Consumption on MS

Worsening of MS symptoms is commonly said to occur due to increased circulating pro-inflammatory factors (Rubin et al. 2020). Evidence of iron deposits in MS lesions was seen in patients with high red meat consumption. Red meat and cow's milk consumption have also been associated with the development of autoantibodies (Riccio and Rossano 2015). Red meat and cow's milk contain the sialic acid N-glycolylneuraminic acid, which is not endogenously formed in humans. The presence of this sialic acid is problematic as humans have a circulating anti-N-glycolylneuraminic acid antibody, which is said to contribute to a state of chronic inflammation upon the consumption of red meat and cow's milk (Riccio and Rossano 2015). However, research on red meat consumption and MS severity, which relies on surveys investigating dietary patterns and symptom severity, is heavily contradictory. What many studies lack is an analysis and correlation of magnetic resonance imaging (MRI) results with dietary choices, as MRI results provide researchers with an objective measure of the extent of demyelination. Studies in Europe noted an association between red meat consumption and worsening MS outcomes, while studies from Iran and Australia do not support this association (Black et al. 2019a, b; Coe et al. 2021; Ertaş Öztürk et al. 2023; Loonstra et al. 2023; Rezaeimanesh et al. 2021; Rubin et al. 2020; Skovgaard et al. 2023).

This divide in findings can be attributed to a multitude of factors. The studies employ different methodologies and research strategies and investigate the effect of red meat consumption on MS severity for different durations of time. Additionally, participants within each study follow widely differing diets with variations in the amount of fruits, vegetables, fats, and milk consumed. Consequently, participants would meet different macro-and micronutrient values. Therefore, from the available evidence, it becomes increasingly difficult to precisely isolate the nutrients with the highest impact on MS severity. Along with the variations in dietary habits, participants may engage in different lifestyle habits, which further influence the course of MS. Genetic variations among populations also play a key role in MS severity, which can lead to differing results on the impact of red meat consumption on MS (Patsopoulos 2018).

A similar dilemma is seen in research on cow's milk consumption and MS severity. Initially, epidemiological research found an association between cow's milk consumption and MS (Kotzamani et al. 2012; Malosse et al. 1993). It was then proposed that the development of antibodies against or an allergy to cow's milk was a risk factor for MS (Chunder et al. 2023). Yet research emerged contradicting these claims, which found no significant difference in antibodies against milk (Ashtari et al. 2013; Monetini et al. 2002; Ramagopalan et al. 2010). In fact, conflicting research has also demonstrated the beneficial effect of cow's milk consumption on MS outcomes (Dehghan and Ghaedi-Heidari 2018; Uygun Özel et al. 2023). Despite these differences, most research demonstrates a negative effect of milk consumption on MS severity.

Research on mice aimed to elucidate the components in milk that may lead to the development of MS. One study first attempted to determine particular proteins in milk found in participants' MS lesions (Winer et al. 2001). They found a heightened T cell response upon exposure to Bovine Serum Albumin (BSA), specifically, the BSA193 epitope, targeted by the immune system. In studies on mice, it was found that a mere 8 of the 29 mice injected with BSA193 developed experimental autoimmune encephalomyelitis (EAE), an animal model for MS, only under the standard EAE-induction protocol (Winer et al. 2001). In another study investigating reactivity to milk protein in pediatric populations of MS, the researchers also found a similar heightened reactivity to the BSA193 epitope (Banwell et al. 2008). Research has shown that butyrophilin, one of the components in milk, mimics myelin in the brain and can lead to encephalitis (Winer et al. 2001). More specifically, it was revealed that the N-terminal domain showed similarities with the human myelin oligodendrocyte glycoprotein, which forms part of axon fibers' lipid-rich coat (Eichinger et al. 2021). However, a study refuted this, claiming that butyrophilin yields protective effects and prevents MS development (Mana et al. 2004). Recently, evidence has highlighted the presence of DNA molecules from cow's milk and red meat in MS lesions (Eilebrecht et al. 2018; Whitley et al. 2014). It was found that these DNA isolates resembled those responsible for transmissible spongiform encephalopathy, a neurodegenerative disease where the accumulation of prion proteins gives the brain a spongy appearance. Several DNA molecules were found, but only two molecules were pursued for research: CMI1.252 and MSBI1.176 (Eilebrecht et al. 2018). These molecules were analyzed in human HEK293TT cells. In these cells, they were significantly transcribed, and the persistent presence of these molecules relied on an interaction between MSBI1.176 and CMI1.252. These molecules were found to be implicated in processes concerning cell viability, progression, and proliferation, which could contribute to the development of MS (Eilebrecht et al. 2018).

Future research endeavors should adopt a holistic approach to decipher the multifaceted relationship between diet and MS severity. This involves not only conducting meticulous dietary surveys but also integrating detailed MRI analyses to provide a more nuanced understanding of the underlying pathology of MS, which could incur prohibitive costs. Additionally, participants should be meticulously matched for various factors, including genetic predispositions, dietary habits, and lifestyle choices, offering a more comprehensive view of how specific factors intricately influence the complex landscape of MS. Such a nuanced approach holds the potential to yield valuable insights, paving the way for the development of personalized dietary interventions tailored to the unique needs of individuals navigating the challenges posed by this complex neurological condition.

6.4 Clinical Trials on Dietary Interventions for MS Management

The potential impact of red meat and milk consumption on MS has been the subject of considerable investigation. One proposed model posits a complex interaction involving latent bovine milk factors, reactivated herpesvirus DNA, and vitamin D deficiency in MS development (Zur Hausen and de Villiers 2015). Furthermore, studies have isolated bovine meat and milk factors (BMMF) from dairy products, in which a high prevalence of antibodies targeting the replication protein of BMMF was detected in brain and serum samples derived from individuals with MS (Zur Hausen et al. 2017). These findings suggest that BMMF infections, in conjunction with other factors such as herpesvirus infections and vitamin D deficiency, may synergistically contribute to the onset of MS.

Red meat consumption presents several additional potential mechanisms that could affect MS. Its elevated iron content, notably heme iron, may induce nitrosylation and DNA damage, potentially exacerbating MS symptoms (Riccio and Rossano 2015). In a separate study, a decrease in red meat consumption and increased fish consumption led to a reduction in endogenous nitrosylation, specifically in the formation of mutagenic NOCs (Joosen et al. 2010). However, they also found that altering meat and fish intake did not have a significant impact on inflammation (Joosen et al. 2010). Therefore, the role of meat in inducing inflammation in multiple sclerosis remains uncertain.

On the other hand, the presence of abnormal iron deposits at sites of inflammation in MS may suggest the potential role of red meat in exacerbating inflammation (Williams et al. 2012). Additionally, red meat contains Neu5Gc, a sialic acid that humans lack due to an inactivating mutation in the CMAH gene (Padler-Karavani et al. 2008). Incorporating Neu5Gc from dietary sources, particularly red meat, may lead to chronic inflammation due to the presence of circulating anti-Neu5Gc antibodies in humans.

Moreover, the arachidonic acid content in red meat is believed to contribute to the progression of MS (Stenson 2014). Red meat contains arachidonic acid, a precursor of pro-inflammatory eicosanoids such as prostaglandins, thromboxanes, and leukotrienes. Activation of the Th17 pathway by arachidonic acid could potentially contribute to the inflammatory process in MS (Stenson 2014).

Another investigation explored the correlation between red meat and wholegrain bread consumption and plasma biomarkers associated with glucose metabolism, oxidative stress, inflammation, and obesity (Montonen et al. 2013). This cross-sectional analysis utilized data from the European Prospective Investigation into Cancer and Nutrition (EPIC) Potsdam Study, comprising 2198 individuals. The study revealed that high levels of red meat intake correlated with elevated levels of gamma-glutamyltransferase (GGT) and high-sensitivity C-reactive protein (hs-CRP). However, upon adjusting for body mass index (BMI) and waist circumference, the association with hs-CRP lost significance, indicating a potential mediation by obesity. Additionally, red meat consumption was linked to heightened inflammation and oxidative stress due to its rich composition of saturated fat, protein, heme-iron, nitrates, and other compounds (Montonen et al. 2013). Overall, these mechanisms suggest that consuming red meat may have adverse effects on MS by promoting inflammation, DNA damage, and iron accumulation, which are all implicated in the pathogenesis and progression of the disease. However, it is essential to note that more research is needed to fully understand the relationship between red meat consumption and MS.

Consumption of red meat seems to be linked to elevated MS-related symptom scores and fatigue severity among individuals with MS (Ertaş Öztürk et al. 2023). Specifically, those who consumed more than one serving of red meat or its products per day reported elevated MS-related symptom scores, indicating a potential exacerbation of MS symptoms with increased red meat intake. Conversely, a noteworthy trend was observed in participants consuming three or more servings of fish per week, showing lower symptom scores and suggesting a possible beneficial effect of fish consumption.

As discussed in a subsequent chapter of this book, reducing red meat consumption is in line with the Mediterranean diet, known for its health benefits and reduced risk of MS. Studies have proposed that limited red meat consumption, as part of the Mediterranean diet, may offer advantages to individuals at high risk of MS. Moreover, intervention studies implementing a modified Mediterranean diet, emphasizing fruits, vegetables, fish, nuts, legumes, and whole grains while restricting red meat intake, resulted in decreased fatigue, MS symptoms, and disability (Ertaş Öztürk et al. 2023). This association between reduced fatigue severity and adherence to a Mediterranean diet may be related to its anti-inflammatory properties. By reducing the intake of inflammatory components like saturated fatty acids found in red meat and increasing anti-inflammatory choices, such as fruits, vegetables, whole grains, and fish, MS symptoms may be positively influenced.

A cross-sectional study in Iran compared dietary patterns among women diagnosed with relapsing-remitting multiple sclerosis (RRMS) to healthy controls (Jahromi et al. 2012). Researchers identified seven major dietary patterns using a validated 168-item Food Frequency Questionnaire (FFQ) and factor analysis. Logistic regression analysis revealed an intriguing inverse correlation between traditional, vegetarian-like, and lacto-vegetarian-like dietary patterns and the risk of

MS. At the same time, a positive association was observed with a high animal fat dietary pattern. The study shed light on the significant impact of red meat consumption on MS risk. Several major dietary patterns emerged through factor analysis, including a traditional pattern characterized by high consumption of low-fat dairy products, red meats, vegetable oils, whole grains, and legumes. Remarkably, this traditional pattern exhibited an inverse association with MS risk, suggesting a potential protective effect against the development of the disease (Jahromi et al. 2012). Conversely, a dietary pattern high in animal fats, likely encompassing red meat among other sources, demonstrated a positive association with MS risk.

In a cross-sectional study involving 2410 persons with multiple sclerosis (pwMS) compared to 24,852 controls, dietary data using a food frequency questionnaire were examined (Coe et al. 2021). The study aimed to assess the consumption of nutrients, focusing on anti-inflammatory food groups like carotene, magnesium, oily fish, and fruits and vegetables, as well as pro-inflammatory nutrients such as saturated fat, sodium, sucrose, red meat, and high-fat dairy products among pwMS, and its correlation with fatigue and quality of life. Analysis revealed significantly lower nutrient intake of all food groups in the MS group compared to controls ($P < 0.05$). Specifically, fish consumption was correlated with lower clinical fatigue ($P < 0.05$). Positive health outcomes on quality-of-life measures were also linked to higher intake of carotene, magnesium, oily fish, fruits, vegetables, and sodium. In contrast, intake of fiber, red meat, and saturated fat (in women) was associated with worse outcomes ($P < 0.05$) (Coe et al. 2021). These findings suggest that pwMS exhibit altered dietary patterns compared to controls, potentially influencing symptom severity. Thus, it emphasizes the importance of dietary interventions in managing MS symptoms and improving quality of life.

Moreover, adherence to a plant-rich diet, characterized by high vegetable intake, was found to be associated with reduced symptom burden across various MS-related symptoms compared to the Western dietary cluster, which had a high intake of red meat (Skovgaard et al. 2023). Specifically, the plant-rich dietary cluster exhibited a reduction in symptom burden for pain, bladder dysfunction, and overall MS-related symptoms. Notably, a high intake of vegetables was associated with a significant reduction in symptom burden compared to low levels of vegetable intake. Interestingly, the study also revealed increased muscle weakness in both high and low meat intake groups but reduced muscle weakness with medium levels of meat intake (Skovgaard et al. 2023).

Another study focusing on pwMS revealed that those who frequently consumed red meat exhibited significantly higher lesion volumes compared to those with lower red meat consumption (Loonstra et al. 2023). Several factors may contribute to this effect. Firstly, red meat is rich in saturated fat and cholesterol, which are known to contribute to vascular dysfunction and inflammation. Additionally, the advanced glycation endproducts formed during the cooking process of red meat can induce oxidative stress and inflammation. Moreover, the high levels of heme iron present in red meat may lead to iron-mediated oxidative stress, further exacerbating the situation (Loonstra et al. 2023).

Pourtostadi et al. found that red meat intake among patients with MS was significantly higher than in healthy controls ($p < 0.001$) (Pourostadi et al. 2021). This suggests that excessive consumption of red meat might act as a predisposing factor for MS. One plausible explanation is related to vitamin D deficiency, which plays a crucial role in the development of MS. As red meat is not a significant source of vitamin D, it can be argued that it may indirectly contribute to vitamin D deficiency, thus indirectly contributing to elevating the risk of developing MS. Excessive consumption of red meat, especially processed red meat, has also been linked to chronic inflammation in the body. This mechanism is heavily implicated in the pathogenesis and progression of MS. Moreover, red meat consumption has been associated with alterations in gut microbiota composition (Pourostadi et al. 2021) or dysbiosis of gut microbiota, which has been implicated in various autoimmune diseases, including MS. Therefore, it is plausible that the changes in gut microbiota induced by red meat consumption may influence immune responses and contribute to MS pathogenesis. Dysbiosis is discussed at length in Chap. 3 of this book. Overall, the study suggests that excessive red meat consumption may act as a predisposing factor for MS, possibly through mechanisms involving vitamin D deficiency, inflammation, and alterations in gut microbiota.

In contrast, a cohort study did not find a statistically significant association between high meat intake (both red and processed) and the risk of late-onset chronic inflammatory diseases (CIDs) (Rubin et al. 2020), with adjusted hazard ratios for developing CID for medium meat intake and high meat intake (compared to low meat intake as a reference) at 0.96 [0.86–1.09] and 0.94 [0.82–1.07], respectively.

Another study delved into the role of dietary factors during adolescence in the risk of primary progressive multiple sclerosis (PPMS) (Rezaeimanesh et al. 2021). This investigation revealed significant insights into the effect of meat intake on PPMS risk. Firstly, it uncovered a substantial inverse association between higher consumption of red meat and poultry during adolescence and the risk of PPMS. This association demonstrated a dose-dependent relationship, with increased intake correlating with a lower risk of PPMS. Specifically, individuals with higher red meat and poultry intake, particularly those in the third tertile, exhibited a notable reduction in PPMS risk, even after adjustments in the fully adjusted model. Furthermore, beyond red meat and poultry, other food groups such as dairy, seafood, vegetables, fruits, and nuts also exhibited a significant inverse association with PPMS risk when consumed at higher levels during adolescence. Additionally, nutrient supplementation with calcium, iron, folic acid, vitamin B12, and vitamin C during adolescence displayed a remarkable association with a substantial decrease of over 84% in PPMS risk (Rezaeimanesh et al. 2021). This underscores the potential role of adequate nutrient intake during adolescence in mitigating the risk of adult-onset PPMS. In conclusion, the study suggests that dietary habits during adolescence, encompassing increased consumption of red meat, poultry, various food groups, and nutrient supplementation, may contribute to reducing the risk of PPMS in adulthood. These findings underscore the importance of dietary practices and nutrient intake during adolescence in shaping the risk of developing PPMS later in life, although further research is warranted to validate these associations and elucidate the underlying mechanisms.

In a study by Black et al., a one standard deviation increase in non-processed red meat density was associated with a 19% reduced risk of a first clinical diagnosis (FCD) of central nervous system demyelination (Black et al. 2019b). This association was significant in females, with a 26% reduced risk of FCD per 22 g/1000 kcal/day increase in non-processed red meat density, but not in males. Adjustment for very long-chain omega-3 polyunsaturated fatty acid (VLCn3PUFA) density did not change the association, suggesting an independent effect of non-processed red meat consumption (Black et al. 2019b). There was no statistically significant association between processed red meat density and the risk of FCD in the total population or when stratified by sex. The association between non-processed red meat consumption and FCD risk was observed only in females, possibly due to the higher prevalence of MS in females. Another hypothesis is that iron sufficiency is important in preventing MS, and females have greater iron requirements (Black et al. 2019b).

To examine the link between milk consumption and MS risk, a group of researchers used phenome-wide Mendelian randomization analysis (MR-PheWAS) and two-sample MR analysis (Yuan et al. 2022). Analyzing data from the UK Biobank involving over 300,000 participants, they discovered that individuals with a genetically predicted inclination towards higher milk consumption—estimated based on their genetic profile—exhibited a reduced risk of MS. This association was confirmed through both MR-PheWAS and two-sample MR analyses, providing robust evidence supporting the idea that milk consumption may act as a protective factor against MS development in the European population (Yuan et al. 2022).

Furthermore, epidemiological studies from various countries have reported a positive correlation between milk consumption and MS prevalence (Morin et al. 2024). For instance, a study in Denmark and the United States has linked dairy intake with increased MS risk. Notably, a Danish study found a doubled risk of MS among dairy workers compared to the general population (Morin et al. 2024). The Nurses' Health Study also identified a positive association between high milk consumption during adolescence and MS development later in life. Additionally, molecular studies have suggested mechanisms by which milk consumption might influence MS pathology, including the role of specific milk antigens in triggering immune responses that target central nervous system antigens (Chunder et al. 2023). Specifically, patients with MS were found to exhibit a significantly higher IgG response to cow milk ($P = 0.0002$) and goat milk ($P = 0.0004$) compared to healthy donors. Moreover, correlation analysis suggested a relationship between the presence of antibodies to cow milk proteins and cross-reactive CNS antigens, particularly myelin oligodendrocyte glycoprotein (MOG) and myelin-associated glycoprotein (MAG), which are known to be targeted in MS. This suggests a possible mechanism of molecular mimicry where antibodies developed against cow milk antigens cross-react with CNS antigens, potentially exacerbating MS pathology (Chunder et al. 2023).

In another study, researchers investigated the genetic differences between monozygotic twins who are discordant for various clinical phenotypes (Vadgama et al.

2019). Among the phenotypes explored was lactase non-persistence (lactose intolerance), and the study identified a de novo variant in the *PLCB1* gene associated with this condition. *PLCB1* plays a role in lipid phosphorus hydrolysis in milk from dairy cows, suggesting a potential link between milk consumption and lactase non-persistence. The study also detected a somatic frameshift deletion in the *PLCB1* gene in the affected twin with lactase non-persistence, providing evidence of a genetic variant in *PLCB1* associated with this condition, which could implicate milk consumption (Vadgama et al. 2019).

Examining the impact of milk, specifically bovine β-casein, on MS, another study measured antibodies to bovine β-casein across various immune-and non-immune-mediated diseases, including MS (Monetini et al. 2002). They discovered a notable rise in antibody levels to beta-casein in Type 1 diabetes, coeliac disease, and latent autoimmune diabetes in adults (LADA) compared to age-matched controls ($p = 0.01$, $p = 0.02$, and $p = 0.01$, respectively). Conversely, individuals with Type 2 diabetes exhibited considerably lower levels of antibodies to beta-casein. This observation might be due to the compromised immune response in Type 2 diabetes patients, particularly those with prolonged duration and inadequate metabolic regulation. Moreover, results indicated no significant differences in β-casein antibody levels between MS patients and age-matched controls, suggesting no elevated antibody response to β-casein in MS patients compared to controls. Therefore, based on the study's findings, there is no compelling evidence to suggest a significant effect of milk, specifically bovine β-casein, on multiple sclerosis (Monetini et al. 2002).

In summary, research into the impact of red meat and dairy consumption on MS reveals complex interactions and potential mechanisms. Red meat consumption, attributed to factors like heme iron content and advanced glycation endproducts, may exacerbate inflammation and oxidative stress, contributing to MS. Conversely, evidence on dairy's effect on MS is inconclusive, with some studies suggesting protection while others highlight immune-triggering mechanisms. Dietary patterns like the Mediterranean diet show promise in mitigating MS symptoms, underlining the importance of dietary interventions. Adolescent dietary habits, including high red meat intake, may mitigate MS risk, emphasizing early dietary choices' significance. While some studies hint at milk's protective role against MS, others suggest a positive correlation, necessitating further research. Understanding diet's role in MS could offer insights into disease management and prevention.

6.5 Conclusion

In conclusion, the intricate relationship between red meat and dairy consumption and their implications in MS presents a complex landscape. While red meat offers essential nutrients like protein, vitamins, and minerals, excessive intake, particularly of processed varieties, has been linked to various health risks, including cardiovascular diseases and certain cancers. Similarly, cow's milk provides a range of nutrients crucial for overall health. Yet, research suggests a potential association between milk consumption and autoimmune diseases like MS. Studies exploring the impact of red meat and dairy on MS severity have yielded conflicting results, influenced by factors such as genetic predispositions, variations in dietary habits, and lifestyle choices among participants. Emerging evidence suggests potential mechanisms through which red meat and dairy intake may exacerbate inflammation, oxidative stress, and autoimmune responses, all implicated in MS progression. Despite these uncertainties, dietary interventions like the Mediterranean diet, emphasizing plant-based foods and limiting red meat intake, have shown promise in reducing MS symptoms and disability. Moving forward, a comprehensive approach integrating dietary surveys with advanced imaging techniques is crucial to unraveling the intricate interplay between diet and MS outcomes, paving the way for personalized dietary interventions tailored to individual needs and optimizing health outcomes for those affected by this complex neurological condition.

6.6 Summary

This chapter investigates the nutritional value and health implications of red meat and cow's milk, particularly in relation to MS. Red meat, rich in protein, fats, and essential micronutrients, is associated with cardiometabolic diseases and cancer due to carcinogens formed during processing and cooking. Cow's milk, while providing essential nutrients, is linked to autoimmune diseases like MS, potentially exacerbating inflammation and oxidative stress. Preclinical evidence shows a complex interaction between dietary factors and MS severity, with red meat consumption implicated in iron deposition in MS lesions and autoantibody development. Clinical trials suggest that reducing red meat intake and adopting a Mediterranean diet may alleviate MS symptoms, though studies on cow's milk consumption yield mixed results. Dietary interventions, including plant-rich diets, show promise for managing MS symptoms. However, further research is needed to develop personalized dietary strategies, particularly considering the potential impact of dietary habits during adolescence on MS risk.

References

Ashtari F, Jamshidi F, Shoormasti RS, Pourpak Z, Akbari M (2013) Cow's milk allergy in multiple sclerosis patients. J Res Med Sci 18(Suppl 1):S62

Banwell B, Bar-Or A, Cheung R, Kennedy J, Krupp LB, Becker DJ, Dosch HM (2008) Abnormal T-cell reactivities in childhood inflammatory demyelinating disease and type 1 diabetes. Ann Neurol 63(1):98–111

Black LJ, Baker K, Ponsonby A-L, Van Der Mei I, Lucas RM, Pereira G, Group, A. I (2019a) A higher mediterranean diet score, including unprocessed red meat, is associated with reduced risk of central nervous system demyelination in a case-control study of Australian adults. J Nutr 149(8):1385–1392

Black LJ, Bowe GS, Pereira G, Lucas RM, van der Mei I, Sherriff JL (2019b) Higher non-processed red meat consumption is associated with a reduced risk of central nervous system demyelination. Front Neurol 10:414582

Chunder R, Heider T, Kuerten S (2023) The prevalence of IgG antibodies against milk and milk antigens in patients with multiple sclerosis. Front Immunol 14:1202006. https://doi.org/10.3389/fimmu.2023.1202006

Coe S, Tektonidis TG, Coverdale C, Penny S, Collett J, Chu BT, Dawes H (2021) A cross sectional assessment of nutrient intake and the association of the inflammatory properties of nutrients and foods with symptom severity in a large cohort from the UK multiple sclerosis registry. Nutr Res 85:31–39

Correale J (2021) Immunosuppressive amino-acid catabolizing enzymes in multiple sclerosis. Front Immunol 11:600428

Dehghan M, Ghaedi-Heidari F (2018) Environmental risk factors for multiple sclerosis: a case-control study in Kerman, Iran. Ir J Nurs Midwif Res 23(6):431–436

Eichinger A, Neumaier I, Skerra A (2021) The extracellular region of bovine milk butyrophilin exhibits closer structural similarity to human myelin oligodendrocyte glycoprotein than to immunological BTN family receptors. Biol Chem 402(10):1187–1202

Eilebrecht S, Hotz-Wagenblatt A, Sarachaga V, Burk A, Falida K, Chakraborty D, Sauerland C (2018) Expression and replication of virus-like circular DNA in human cells. Sci Rep 8(1):2851

Ertaş Öztürk Y, Helvaci EM, Sökülmez Kaya P, Terzi M (2023) Is Mediterranean diet associated with multiple sclerosis related symptoms and fatigue severity? Nutr Neurosci 26(3):228–234. https://doi.org/10.1080/1028415X.2022.2034241

Farvid MS, Sidahmed E, Spence ND, Mante Angua K, Rosner BA, Barnett JB (2021) Consumption of red meat and processed meat and cancer incidence: a systematic review and meta-analysis of prospective studies. Eur J Epidemiol 36(9):937–951. https://doi.org/10.1007/s10654-021-00741-9

Godfray HCJ, Aveyard P, Garnett T, Hall JW, Key TJ, Lorimer J, Jebb SA (2018) Meat consumption, health, and the environment. Science 361(6399):eaam5324. https://doi.org/10.1126/science.aam5324

Harirchian M, Bitarafan S, Honarvar N (2016) Dairy products consumption in multiple sclerosis patients: useful or harmful. Int J Neurorehab 3:e126

Haug A, Høstmark AT, Harstad OM (2007) Bovine milk in human nutrition—a review. Lipids Health Dis 6(1):25. https://doi.org/10.1186/1476-511X-6-25

Herforth A, Arimond M, Álvarez-Sánchez C, Coates J, Christianson K, Muehlhoff E (2019) A global review of food-based dietary guidelines. Adv Nutr 10(4):590–605. https://doi.org/10.1093/advances/nmy130

Hess JM, Stephensen CB, Kratz M, Bolling BW (2021) Exploring the links between diet and inflammation: dairy foods as case studies. Adv Nutr 12:1S–13S. https://doi.org/10.1093/advances/nmab108

Hobbs-Grimmer DA, Givens DI, Lovegrove JA (2021) Associations between red meat, processed red meat and total red and processed red meat consumption, nutritional adequacy and markers of health and cardio-metabolic diseases in British adults: a cross-sectional analysis using data from UK National Diet and nutrition survey. Eur J Nutr 60(6):2979–2997. https://doi.org/10.1007/s00394-021-02486-3

Jahromi SR, Toghae M, Jahromi MJR, Aloosh M (2012) Dietary pattern and risk of multiple sclerosis. Ir J Neurol 11(2):47

Joosen AM, Lecommandeur E, Kuhnle GG, Aspinall SM, Kap L, Rodwell SA (2010) Effect of dietary meat and fish on endogenous nitrosation, inflammation and genotoxicity of faecal water. Mutagenesis 25(3):243–247

Kotzamani D, Panou T, Mastorodemos V, Tzagournissakis M, Nikolakaki H, Spanaki C, Plaitakis A (2012) Rising incidence of multiple sclerosis in females associated with urbanization. Neurology 78(22):1728–1735. https://doi.org/10.1212/WNL.0b013e31825830a9

Loonstra FC, de Ruiter LR, Schoonheim MM, Moraal B, Strijbis EM, de Jong BA, Uitdehaag BM (2023) The role of diet in multiple sclerosis onset and course: results from a nationwide retrospective birth-year cohort. Ann Clin Trans Neurol 10(8):1268–1283

Malosse D, Perron H (1993) Correlation analysis between bovine populations, other farm animals, house pets, and multiple sclerosis prevalence. Neuroepidemiology 12(1):15–27. https://doi.org/10.1159/000110295

Malosse D, Perron H, Sasco A, Seigneurin JM (1993) Correlation between Milk and dairy product consumption and multiple sclerosis prevalence: a worldwide study. Neuroepidemiology 11(4–6):304–312. https://doi.org/10.1159/000110946

Mana P, Goodyear M, Bernard C, Tomioka R, Freire-Garabal M, Linares D (2004) Tolerance induction by molecular mimicry: prevention and suppression of experimental autoimmune encephalomyelitis with the milk protein butyrophilin. Int Immunol 16(3):489–499

Marangoni F, Pellegrino L, Verduci E, Ghiselli A, Bernabei R, Calvani R, Piretta L (2019) Cow's milk consumption and health: a health professional's guide. J Am Coll Nutr 38(3):197–208

Melini F, Melini V, Luziatelli F, Ruzzi M (2017) Raw and heat-treated milk: from public health risks to nutritional quality. Beverages 3(4):54

Micha R, Michas G, Mozaffarian D (2012) Unprocessed red and processed meats and risk of coronary artery disease and type 2 diabetes—an updated review of the evidence. Curr Atheroscler Rep 14(6):515–524. https://doi.org/10.1007/s11883-012-0282-8

Mladěnka P, Macáková K, Kujovská Krčmová L, Javorská L, Mrštná K, Carazo A, Collaborators (2022) Vitamin K—sources, physiological role, kinetics, deficiency, detection, therapeutic use, and toxicity. Nutr Rev 80(4):677–698. https://doi.org/10.1093/nutrit/nuab061

Monetini L, Cavallo MG, Manfrini S, Stefanini L, Picarelli A, Di Tola M, Di Giulio C (2002) Antibodies to bovine beta-casein in diabetes and other autoimmune diseases. Horm Metab Res 34(08):455–459

Montonen J, Boeing H, Fritsche A, Schleicher E, Joost H-G, Schulze MB, Pischon T (2013) Consumption of red meat and whole-grain bread in relation to biomarkers of obesity, inflammation, glucose metabolism and oxidative stress. Eur J Nutr 52:337–345

Morin CR, Baeva M-E, Hollenberg MD, Brain MC (2024) Milk and multiple sclerosis: a possible link? Mult Scler Relat Disord 105477:105477

Padler-Karavani V, Yu H, Cao H, Chokhawala H, Karp F, Varki N, Varki A (2008) Diversity in specificity, abundance, and composition of anti-Neu5Gc antibodies in normal humans: potential implications for disease. Glycobiology 18(10):818–830

Pan L, Chen L, Lv J, Pang Y, Guo Y, Pei P, Li L (2022) Association of red meat consumption, metabolic markers, and risk of cardiovascular diseases. Front Nutr 9:833271. https://doi.org/10.3389/fnut.2022.833271

Parks NE, Jackson-Tarlton CS, Vacchi L, Merdad R, Johnston BC (2020) Dietary interventions for multiple sclerosis-related outcomes. Cochrane Database Syst Rev 2020:CD004192. https://doi.org/10.1002/14651858.CD004192.pub4

Patsopoulos NA (2018) Genetics of multiple sclerosis: an overview and new directions. Cold Spring Harb Perspect Med 8(7):a028951. https://doi.org/10.1101/cshperspect.a028951

Pourostadi M, Sattarpour S, Poor BM, Asgharzadeh M, Kafil HS, Farhoudi M, Rashedi J (2021) Vitamin D receptor gene polymorphism and the risk of multiple sclerosis in the Azeri population of Iran. Endo Metab Immune Disord Drug Targets 21(7):1306–1311

Ramagopalan SV, Dyment DA, Guimond C, Orton S-M, Yee IM, Ebers GC, Sadovnick AD (2010) Childhood cow's milk allergy and the risk of multiple sclerosis: a population based study. J Neurol Sci 291(1):86–88. https://doi.org/10.1016/j.jns.2009.10.021

Rezaeimanesh N, Moghadasi AN, Sahraian MA, Eskandarieh S (2021) Dietary risk factors of primary progressive multiple sclerosis: a population-based case-control study. Mult Scler Relat Disord 56:103233

Riccio P, Rossano R (2015) Nutrition facts in multiple sclerosis. ASN Neuro 7(1):1759091414568185. https://doi.org/10.1177/1759091414568185

Rodop BB, Sarı E, Erbaş O (2021) Cow's milk and autoimmunity. J Exp Basic Med Sci 2(3):302–307

Rubin KH, Rasmussen NF, Petersen I, Kopp TI, Stenager E, Magyari M, Andersen V (2020) Intake of dietary fibre, red and processed meat and risk of late-onset chronic inflammatory diseases: a prospective Danish study on the "diet, cancer and health" cohort. Int J Med Sci 17(16):2487

Skovgaard L, Trénel P, Westergaard K, Knudsen AK (2023) Dietary patterns and their associations with symptom levels among people with multiple sclerosis: a real-world digital study. Neurol Ther 12(4):1335–1357. https://doi.org/10.1007/s40120-023-00505-5

Stenson WF (2014) The universe of arachidonic acid metabolites in inflammatory bowel disease: can we tell the good from the bad? Curr Opin Gastroenterol 30(4):347–351

Temperley IA, Seldon AN, Reckord MA, Yarad CA, Islam FT, Duncanson K, Maltby VE (2023) Dairy and gluten in disease activity in multiple sclerosis. Mult Scler J Exp Transl Clin 9(4):20552173231218107

Thorning TK, Raben A, Thorning T, Soedamah-Muthu SS, Givens I, Astrup A (2016) Milk and dairy products: good or bad for human health? An assessment of the totality of scientific evidence. Food Nutr Res 60:32527. https://doi.org/10.3402/fnr.v60.32527

Uygun Özel S, Bayram S, Kılınç M (2023) The relationship between dietary profile and adherence to the Mediterranean diet with EDSS and quality of life in multiple sclerosis patients: a retrospective cross-sectional study. Nutr Neurosci 27:1–9

Vadgama N, Pittman A, Simpson M, Nirmalananthan N, Murray R, Yoshikawa T, Hughes D (2019) De novo single-nucleotide and copy number variation in discordant monozygotic twins reveals disease-related genes. Eur J Hum Genet 27(7):1121–1133

Vojdani A (2014) A potential link between environmental triggers and autoimmunity. Autoimmune Dis 2014:437231. https://doi.org/10.1155/2014/437231

Whitley C, Gunst K, Müller H, Funk M, Zur Hausen H, de Villiers E-M (2014) Novel replication-competent circular DNA molecules from healthy cattle serum and milk and multiple sclerosis-affected human brain tissue. Genome Announc 2(4):00849–00814. https://doi.org/10.1128/genomea

Williams P (2007) Nutritional composition of red meat. Nutr Dietet 64:S113–S119

Williams R, Buchheit CL, Berman NE, LeVine SM (2012) Pathogenic implications of iron accumulation in multiple sclerosis. J Neurochem 120(1):7–25

Williamson C, Foster R, Stanner S, Buttriss J (2005) Red meat in the diet. Nutr Bull 30(4):323–355

Winer S, Astsaturov I, Cheung RK, Schrade K, Gunaratnam L, Wood DD, Becker DJ (2001) T cells of multiple sclerosis patients target a common environmental peptide that causes encephalitis in mice. J Immunol 166(7):4751–4756

Yuan S, Sun J, Lu Y, Xu F, Li D, Jiang F, Larsson SC (2022) Health effects of milk consumption: phenome-wide Mendelian randomization study. BMC Med 20(1):455

Zur Hausen H, de Villiers EM (2015) Dairy cattle serum and milk factors contributing to the risk of colon and breast cancers. Int J Cancer 137(4):959–967

Zur Hausen H, Bund T, de Villiers E-M (2017) Infectious agents in bovine red meat and milk and their potential role in cancer and other chronic diseases. In: Viruses, genes, and cancer. Springer, Cham, pp 83–116

Chapter 7
Role of Vitamins in Multiple Sclerosis

Haia M. R. Abdulsamad, Amna Baig, Sara Aljoudi, Nadia Rabeh, Zakia Dimassi, and Hamdan Hamdan

Abstract Various types of diets have been investigated for their role in improving or exacerbating the status of Multiple Sclerosis. Analyzing individual components of each diet is essential to understand better how these diets affect MS. Diets high in fibers and certain dietary supplements or vitamins optimize the dietary regimen for MS patients. Indeed, fiber-rich diets are associated with improved cognitive function, which is mediated by the diet's beneficial alterations to the gut microbiota and inflammatory response. Certain over-the-counter (OTC) dietary supplements, such as caffeine, Omega-3, and Melatonin, potentially improve MS status or support MS therapy. Vitamins are the most extensively studied nutritional components. In particular, Vitamin D has a clinically proven role in MS pathogenesis and modulation. Other water-and fat-soluble vitamins are still under investigation and are showing promising results in inflammation, motor symptoms, and psychological manifestations of MS.

Keywords Multiple Sclerosis (MS) · Vitamins · Ritinoic Acid · Thiamin · Folic Acid · Experimental autoimmune encephalomyelitis (EAE)

H. M. R. Abdulsamad · A. Baig · S. Aljoudi · N. Rabeh · H. Hamdan (✉)
Department of Biological Sciences, College of Medicine and Health Sciences, Khalifa University, Abu Dhabi, United Arab Emirates
e-mail: hamdan.hamdan@ku.ac.ae

Z. Dimassi
Department of Medical Sciences, College of Medicine and Health Sciences, Khalifa University, Abu Dhabi, United Arab Emirates

© The Author(s), under exclusive license to Springer Nature Singapore Pte Ltd. 2024
H. Hamdan (ed.), *Exploring the Effects of Diet on the Development and Prognosis of Multiple Sclerosis (MS)*, Nutritional Neurosciences, https://doi.org/10.1007/978-981-97-4673-6_7

Abbreviations

BBB	Blood-brain barrier
Biotin	Vitamin B7
CCL2	Chemokine CeC motif ligand 2
Cobalamin	Vitamin B12
CRP	C-reactive protein
EAE	Experimental autoimmune encephalomyelitis
EDSS	Expanded Disability Status Scale
FBS	Fasting blood sugar
Folic acid	Vitamin B9
FoxP3	Forkhead box P3
Gas6	Growth arrest specific gene 6
Gd	Gadolinium
IFN-β-1b	Interferon-β-1b
IL-10	Interleukin-10
MS	Multiple Sclerosis
MUFAs	Mono-unsaturated fatty acids
NAD	Nicotinamide adenine dinucleotide
NADP	Nicotinamide adenine dinucleotide phosphate
NF-κB	Nuclear factor kappa B
Niacin	Vitamin B3
Nrf2	Nuclear E2-related factor
ON	Optic neuritis
OTC	Over-the-counter
PUFAs	Polyunsaturated fatty acids
Pyridoxine	Vitamin B6
RA	Retinoic acid
RALDH2	Retinaldehyde dehydrogenase 2
RARα-γ	Retinoic acid receptors
Retinoic Acid	Vitamin A
RORα-γ	Retinoic acid orphan receptors
ROS	Reactive oxygen species
RXRα-γ	Retinoid X receptors
RXR-γ	Retinoid X receptor gamma
Th1	T helper 1
Th17	T helper 17
Thiamine	Vitamin B1
TLRs	Toll-like receptors
TNF-α	Tumor necrosis factor-alpha
Treg	T regulatory
VCAM-1	Vascular cell adhesion molecule

Learning Objectives
- Identify the roles of high-fiber diets, vitamins, and dietary supplements in managing MS.
- Explain how different vitamins, such as A, D, E, K, and B-complex, influence the pathogenesis and management of MS.
- Assess the potential benefits of over-the-counter dietary supplements like caffeine, Omega-3, and melatonin in supporting MS therapy.
- Investigate the biochemical and immunological mechanisms through which dietary components affect MS, including gut microbiota and inflammatory responses.
- Discuss the safety concerns and evidence gaps in using various vitamins and supplements for MS treatment.

7.1 Introduction

For decades, Disease-modifying therapy has been the cornerstone of MS treatment. This approach focuses on restoring motor function, preventing disease relapse, and maintaining remission. To date, the optimal lifestyle and diet for MS patients continue to be an area of intensive research, as these factors affect the disease itself along with its comorbidities. The various dietary and nutritional interventions, such as vitamins, supplements and high-fibers for the treatment of MS will be explored in chapters 7–9 and 11.

7.2 Vitamins

Vitamins play a crucial role in various biochemical mechanisms within our cells, from gene expression to modulating signaling pathways. This section illustrates the role of vitamins in MS. Through changes in gene expression, vitamins, namely B1, B6, B3, D, and E, can influence cells of the immune system, such as T helper 1 (Th1) and T helper 17 (Th17), to induce an anti-inflammatory state (Fanara et al. 2021; Orti et al. 2020). Furthermore, vitamin A can improve the psychophysiological symptoms in MS patients, like depression and fatigue (Stoiloudis et al. 2022). It is hypothesized that the release of inflammatory cytokines precipitates symptoms such as depression and fatigue in the setting of MS. The release of these cytokines can be inhibited by administering vitamin A, hence its potential use in the treatment of MS (Sama et al. 2016). Given its vital established role in MS, the role of Vitamin D will be discussed in depth in a separate section. A summary of vitamins and their implication in MS treatment can be found in Table 7.1.

Table 7.1 Summary of vitamins' potential effects on MS

	Vitamin	Summary of effects on MS
Fat-soluble vitamins	A	1. Inhibits Th17 cells 2. Regulates oligodendrocyte to aid in remyelination 3. Balances Th1, Th17 and Th19 cells with Th2 and Treg 4. Alters B cells and dendritic cell function and enhances tolerance against autoimmunity 5. Increases secretion of IL-10 from B cells 6. Improves the integrity of BBB 7. Reduces *IL17A*, *RORγt* gene expression, IFN-γ, and T-bet levels in peripheral immune cells
	E	1. Inhibits NF-κB
	K	1. Aids in the survival of oligodendrocytes, sphingolipid synthesis and subsequently myelination/ remyelination.
Water-soluble vitamins	C	1. Promotes Fenton's reaction that may worsen the inflammatory status in MS patients
	B1	1. Deficiency can lead to the generation of monocyte chemotactic proteins that activate microglia and T-cell infiltration and increase Th1 and Th17 counts, leading to a deteriorated MS status 2. Improves depression by improving mitochondrial oxidative function[a]
	B3	1. Reduces inflammation in the nervous system and improves EDSS score
	B6	1. Aids in homocysteine uptake, increasing production of cysteine and myelin formation
	B7	1. Improves motor score, muscle strength, fatigue, coordination, and visual acuity 2. May interact with patients' laboratory tests 3. High doses might increase relapse risk and transient myopathy in a patient on high-dose biotin
	B9	1. Increases homocysteine uptake
	B12	1. Increases homocysteine uptake 2. Phospholipids production and remyelination 3. Improves motor symptoms manifestation 4. Improvises EDSS score and relapse risk

[a] A balance diet is needed to produce optimum effect
Expanded Disability Status Scale (EDSS)

7.2.1 Vitamin a (Retinoic Acid)

Vitamin A, a fat-soluble vitamin with a proven role in vision, epithelial differentiation, and growth, has an important role in immune function (Runia et al. 2014). Vitamin A exerts an effect through its active metabolite, retinoic acid (RA). RA helps in immune regulation and neuroplasticity and has been shown to inhibit Th17 cells (Runia et al. 2014). The role of Vitamin A in inflammation and immunity in relation to Th17 and T regulatory (Treg) cells has been extensively researched through in vitro studies. Since a component of MS pathogenesis involves imbalances between Th17 and Treg, vitamin A could have a role in the pathogenesis and management of MS (Tryfonos et al. 2019).

RA is a component involved in mechanisms of immunomodulation. RA has been implicated in the formation of Treg cells that express forkhead box P3 (Foxp3), which ensures activation and represses anti-inflammatory and pro-inflammatory genes, respectively (Runia et al. 2014). RA is a signaling molecule that activates nuclear retinoid receptors in 3 sub-groups: retinoic acid receptors (RARα-γ), retinoic acid orphan receptors (RORα-γ), and retinoid X receptors (RXRα-γ). The role of RXR will be thoroughly discussed in the Omega-3 and vitamin D sections. Through specific retinoid X receptor gamma (RXR-γ), which supports remyelination (Khosravi-Largani et al. 2018), vitamin A acts on retinoid receptors to regulate oligodendrocyte differentiation (Rahimi-Dehgolan et al. 2023). Upon RXR receptor activation, the balance between Th1, Th17, and Th19 cells with Th2 and Treg cells is adjusted. Moreover, its activation leads to changes in B cell and dendritic cell function and enhanced tolerance against autoimmunity. Compared to healthy controls, the B cells in MS patients secreted fewer anti-inflammatory cytokines, namely interleukin-10 (IL-10), when stimulated by toll-like receptors (TLRs) RP105 and TLR9. RA increased the release of IL-10 from MS patients' B cells without affecting the release of pro-inflammatory cytokines such as tumor necrosis factor-alpha (TNF-α). RA's effect on the IL-10:TNF- α ratio was comparable to that of treatments such as glatiramer acetate or interferon-β-1b (IFN- β -1b), indicating vitamin A's potential as a therapeutic strategy for patients with MS (Eriksen et al. 2015). Through the activation of the RXR, vitamin D can potentiate the effects of vitamin A, indicating a therapeutic advantage of concomitant vitamin D and A administration (Riccio and Rossano 2018). The collective evidence thus illustrates the involvement of vitamin A in the pathogenesis and management of MS (Tryfonos et al. 2019).

The loss of the blood-brain barrier (BBB) integrity in lesions of MS is orchestrated by high levels of IL-6, chemokine CeC motif ligand 2 (CCL2), and vascular cell adhesion molecule (VCAM-1). Consequently, an upregulation in the production of RA from astrocytes plays an essential role in restoring the integrity of the BBB in patients with MS. The elevated levels of RA are supported by enhanced expression of retinaldehyde dehydrogenase 2 (RALDH2) and antioxidant transcription factor nuclear E2-related factor (Nrf2), which will decrease the abundance of reactive oxygen species (ROS) in the BBB (Khosravi-Largani et al. 2018).

One randomized controlled study conducted in mouse models found that the consumption of vitamin A can positively affect the course of disease in patients with RRMS. In these mouse models, levels of *IL17A*, *RORγ* isoform *RORγt* gene expression, IFN-γ, and lymphocyte transcription factor T-bet levels in peripheral immune cells decreased, along with interfering with the differentiation of Th17 cells (Evans et al. 2018; Mousavi Nasl-Khameneh et al. 2018). Furthermore, concurrent administration of omega-3 improved the suppression effect of vitamin A on *IL17A* and *RORγt* gene expression in a dose-dependent manner. These two genes have a documented role during relapse (Mousavi Nasl-Khameneh et al. 2018). However, vitamin A consumption did not show any radiologic or clinical signs of improvement (Evans et al. 2018). Another study revealed that supplementation of vitamin A derivative for 6 months improved the functionality of MS patients. Yet there was no effect on the Expanded Disability Status Scale (EDSS) score, a scale from 1 to 10 used to quantify disability in MS patients, or brain T2 active lesions' size and intensity, a diagnostic radiological feature of MS possibly indicating demyelination (Bitarafan et al. 2015; Filippi et al. 2019). Despite these unsuccessful attempts to radiologically probe for improvements in MS with vitamin A administration, a study documented improvements through Gadolinium (Gd) enhancement on MRI imaging. Areas enhanced by Gd contrast are an indicator of BBB disruption and active inflammation. In a study on RRMS patients taking vitamin A subcutaneous injections thrice weekly, it was found that an increase in retinol level by 1 micromole decreases the risk of developing Gd-enhancing lesions by 49%. The study also noted a 49 and 42% decrease in the risk of T2 and active lesions for every 1 micromole increase in retinol (Løken-Amsrud et al. 2013).

One of the major neuro-ophthalmic presentations in MS is optic neuritis (ON), an optic nerve inflammation causing atrophy and axonal loss shown as "*a unilateral acute visual loss with retrobulbar pain exacerbated with eye movement* (Atkins 2014)." Corticosteroid therapy is used for effective short-term management, but currently, therapeutic approaches for long-term management are lacking. One double-blinded placebo-controlled clinical trial investigated the effect of vitamin A administration on ON due to its regenerative and immune modulation functions. The study enrolled 50 patients and revealed that the extent of neuronal loss in patients administered vitamin A was lower during the episode of ON. These beneficial effects were attributed to an increase in IL-10, GATA3, and IL-4 gene expression, as well as a decrease in IL17 gene expression, T-bet expression, and Th17 differentiation. With all the previous mentioned actions, there was no effect on the functional integrity of optical pathways (Rahimi-Dehgolan et al. 2023).

A randomized, blinded clinical trial investigated the effect of supplementation with 25,000 IU/day of vitamin A in 35 MS compared to placebo on C-reactive protein (CRP), liver enzymes, fasting blood sugar (FBS), and lipid profiles. While there was an unexpected prominent increase in CRP, no significant difference was detected in the other parameters. Although 25,000 IU/day of vitamin A has not been shown to cause side effects, the steady increase in CRP warrants cautious vitamin A supplementation with frequent checkups. This is because Vitamin A has a long half-life with rapid absorption and slow clearance from the body due to its fat solubility. Combining these features increases the chances of vitamin A toxicity in the long term and increases serum triglycerides and liver enzymes, hence the importance of additional investigations to probe for any side effects (Jafarirad et al. 2013).

In contrast to the previous findings, a longitudinal case-control study showed no association between relapse rate and vitamin A levels (Runia et al. 2014). Due to conflicting evidence, the therapeutic effect of vitamin A for MS warrants further investigation. Synthesizing the referenced studies, we argue that vitamin A may not protect against MS relapse. Rather, vitamin A can reduce the severity of the attack and aid in the remyelination process through increasing RA levels in myelin debris phagocytosis, an essential component in remyelination pathways, and possibly through its antioxidant effect (Rahimi-Dehgolan et al. 2023). Current evidence on the therapeutic use and safe dosage of vitamin A for MS is lacking. Additional evidence is needed to justify its use in the setting of MS.

7.2.2 Vitamin E

Vitamin E is a fat-soluble vitamin that belongs to tocotrienols and tocopherols. It has multiple activities, such as antioxidation, gene regulation, cell signaling, and immune modulation (Khosravi-Largani et al. 2018). Vitamin E is found to inhibit Nuclear factor kappa B (NF-κB), which is a transcription factor involved in cell apoptosis and cell proliferation. At the same time, in an EAE model, NF-κB showed increased activity that eventually decreased during recovery. From this evidence, it was proposed that since vitamin E inhibits NF-κB, it may be helpful in MS (Khosravi-Largani et al. 2018). To reap the benefits of vitamin E administration, concurrent supplementation with cholesterol and triglycerides seems to influence MS treatment. In a randomized, placebo-controlled, double-blind clinical study, a reduction in relapse rate was seen using a mixture of several polyunsaturated fatty acids (PUFAs), mono-unsaturated fatty acids (MUFAs), saturated fatty acids, and vitamins E and A (Marios et al. 2013). Within this combination, it was noted that vitamin E was a vital component. More importantly, it was found that when administered alone it did not significantly decrease the relapse rate (Khosravi-Largani et al. 2018; Marios et al. 2013; Pahan and Schmid 2000).

7.2.3 Vitamin K

Vitamin K is a fat-soluble vitamin with a well-established role in hemostasis and anti-inflammatory potential. Few studies reported that the vitamin K-dependent growth arrest-specific gene 6 (Gas6) plays a role in MS by contributing to the survival of oligodendrocytes, sphingolipid synthesis, and subsequently, myelination/remyelination. However, the details of the role of Vitamin K in MS pathogenesis are yet to be demonstrated (Binder et al. 2011; Ferland 2012; Khosravi-Largani et al. 2018; Sainaghi et al. 2013).

7.2.4 Vitamin C

Vitamin C is a water-soluble vitamin that has a role in improving cognitive function. Despite its antioxidant activity in the hippocampus, the adverse effects of vitamin C in MS patients outweigh its potential therapeutic capacity. Some studies revealed that a high vitamin C intake might be harmful as it promotes the Fenton reaction. This reaction produces hydroxyl radicals in the brain and spinal cord white matter, exacerbating the existing inflammation in MS (Khosravi-Largani et al. 2018).

7.2.5 Vitamin B1 (Thiamine)

Vitamin B1 is a water-soluble vitamin shown to have a role in myelination in animal models by regulating key metabolic components: the α-Ketoglutarate dehydrogenase complex in the mitochondria and transketolase activities in the cytosol and extracellular matrix. Vitamin B1's regulation of such metabolic processes is imperative for cellular function. A thiamine deficiency can result in mitochondrial dysfunction, which is marked by an increase in oxidative stress. The underlying mechanism for this dysfunction is delineated by the downregulation of the respiratory chain complex secondary to thiamine deficiency, thus affecting the neurons' ability to produce ATP. Consequently, mitochondrial dysfunction can precipitate axonal demyelination and disruption of neurogenesis and the blood-brain (Abdou and Hazell 2015; Campbell and Mahad 2018; Khosravi-Largani et al. 2018; Ortí et al. 2020). Such alterations in mitochondrial function have been suggested to be implicated in the presentation of depression in MS patients.

Given the correlation between vitamin B1 and depression, it was proposed that vitamin B1 improves depression by improving mitochondrial oxidative function and reducing physiological stress, thus reproducing the result of antidepressants without the associated side effects (Fanara et al. 2021; Orti et al. 2020). A pilot study further highlights the importance of thiamine intake in preventing depression in MS patients. The study recruited 51 patients, with an average age of 47 years, who had been diagnosed with MS >6 months and were on treatment with glatiramer acetate, an immunomodulator, and interferon beta, a cytokine. Through a series of questionnaires, the study further supported the correlation between depression and thiamine deficiency. The study highlighted that the benefits of thiamine supplementation for preventing depression could only be reaped by incorporating a balanced diet and adequate intake of high-quality carbohydrates (Ortí et al. 2020). In addition to mitochondrial dysfunction, vitamin B1 deficiency can lead to the deterioration of MS status by modulating inflammatory mechanisms (Ji et al. 2014). Thiamine deficiency can prompt the generation of monocyte chemotactic proteins that promote microglia and T-cells infiltration and increase Th1 and Th17 counts, leading to a deteriorated MS status (Ji et al. 2014).

7.2.6 Vitamin B3 (Niacin)

Vitamin B3 is a water-soluble vitamin used to produce nicotinamide adenine dinucleotide (NAD) and nicotinamide adenine dinucleotide phosphate (NADP) in the human body for hydrogen transfer in metabolic reactions (Khosravi-Largani et al. 2018). Research on the use of niacin for MS is limited, but a paper in 2015 showed that niacin could reduce inflammation in the nervous system (Offermanns and Schwaninger 2015) and potentially improve patients' EDSS scores (Gn et al. 2011).

7.2.7 Vitamin B6 (Pyridoxine)

Vitamin B6 is a water-soluble vitamin involved in the metabolism of lipids, sugars, amino acids, and, most importantly, the synthesis of neurotransmitters. While the therapeutic effect of pyridoxine levels in MS patients is a subject of debate, the fact that it serves as a cofactor in the uptake of homocysteine, a neurotoxic and potentially neurodegenerative agent, makes it an attractive treatment for mitigating the extent of myelin damage (Khosravi-Largani et al. 2018). Along with vitamin B6, vitamins B9 and B12, discussed later in this chapter, have also been found to be the main drivers in the uptake of homocysteine (Khosravi-Largani et al. 2018).

7.2.8 Vitamin B7 (Biotin)

Vitamin B7 is involved in several critical metabolic pathways as a cofactor. A biotin deficiency is associated with neurological conditions, namely neuropathy and encephalopathy, with neuropathy being a common clinical presentation of MS. In aggressive cases of MS, it has been reported that vitamin B7 supplementation improves motor score, muscle strength, fatigue, coordination, and visual acuity. The underlying hypothesis is that increased energy production through fatty acid oxidation aids remyelination (Evans et al. 2018) by activating the rate-limiting enzyme of myelin synthesis, acetyl-CoA carboxylase. Two studies showed that supplementation of biotin in a dose ranging from 100–300 mg daily led to a reduction in EDSS score and slowed progression (Sedel et al. 2015; Tourbah et al. 2016). However, a phase III RCT in 2020 showed that biotin had no therapeutic advantage as no statistically significant difference from the placebo group was found. The study also showed that biotin may interact with patients' laboratory tests, compromising the results and their interpretation (Cree et al. 2020). More importantly, it was found that high doses have the potential to increase relapse risk (Fanara et al. 2021). An additional factor that diminishes the appeal of biotin supplementation is the transient myopathy experienced in patients on a high dose of biotin (Maillart et al. 2019).

However, a notable limitation of the studies cited above is their small sample size. Therefore, further investigation is needed to confirm an association between biotin and MS pathogenesis and elucidate any possible therapeutic roles.

7.2.9 Vitamin B9 (Folic Acid) and B12 (Cobalamin)

Vitamins B9 and B12 are essential elements in methylation reactions during DNA synthesis and repair, the metabolism of some amino acids and fatty acids, and the nervous system's function. Since the 1950s, the role of vitamin B12 has been explored due to its function in the myelination process (Fanara et al. 2021) and its deficiency has been associated with an early onset of MS. This was proposed due to vitamin B12's action on phospholipids production and remyelination as a cofactor in the production of myelin sheet (Evans et al. 2018). Homocysteine levels are elevated in vitamin B9 and B12 deficiency and in cases of MS. An accumulation of homocysteine is concerning, given its neurotoxic and neurodegenerative properties. By maintaining adequate levels of vitamin B12, the potentially negative effects of homocysteine in MS patients can be prevented (Khosravi-Largani et al. 2018).

Administration of vitamin B12 as MS therapy in a mouse model improved motor symptoms and remyelination (Evans et al. 2018). It could enhance EDSS scores while decreasing relapse risk (Khosravi-Largani et al. 2018). Another role of Vitamin B12 in MS is immune system modulation through influencing cytokines such as TNF-γ. However, as with other vitamins, some studies showed no correlation between folic acid and cobalamin levels in MS (Khosravi-Largani et al. 2018).

7.3 Conclusion

After analyzing the data available on vitamin use in MS, there is strong evidence for the use of vitamins such as D, B9, and B12. There are still safety concerns, and more research is needed to fully understand the role of the other vitamins. Vitamin C results are not MS-friendly, and awareness is required about supplementing patients with doses more than the daily need.

7.4 Summary

This chapter explores the influence of different diets, vitamins, and supplements on MS. High-fiber diets, which benefit cognitive function through positive changes in gut microbiota and inflammation, are highlighted. Over-the-counter dietary supplements such as caffeine, Omega-3, and melatonin show potential benefits for MS patients. Among vitamins, Vitamin D has a well-documented role in MS modulation, while other vitamins like A, B1, B6, and B12 offer promising results in managing inflammation, motor symptoms, and psychological manifestations. Despite the potential benefits, there are safety concerns and mixed evidence regarding some vitamins, necessitating further research to develop personalized dietary strategies for MS management.

References

Abdou E, Hazell AS (2015) Thiamine deficiency: an update of pathophysiologic mechanisms and future therapeutic considerations. Neurochem Res 40(2):353–361. https://doi.org/10.1007/s11064-014-1430-z

Atkins EJ (2014) Optic neuritis. In: Aminoff MJ, Daroff RB (eds) Encyclopedia of the neurological sciences, Second edn. Academic Press, Cambridge, pp 681–686. https://doi.org/10.1016/B978-0-12-385157-4.00170-6

Binder MD, Xiao J, Kemper D, Ma GZ, Murray SS, Kilpatrick TJ (2011) Gas6 increases myelination by oligodendrocytes and its deficiency delays recovery following cuprizone-induced demyelination. PLoS One 6(3):e17727. https://doi.org/10.1371/journal.pone.0017727

Bitarafan S, Saboor-Yaraghi A, Sahraian MA, Nafissi S, Togha M, Beladi Moghadam N, Harirchian MH (2015) Impact of vitamin a supplementation on disease progression in patients with multiple sclerosis. Arch Iran Med 18(7):435–440

Campbell G, Mahad DJ (2018) Mitochondrial dysfunction and axon degeneration in progressive multiple sclerosis. FEBS Lett 592(7):1113–1121. https://doi.org/10.1002/1873-3468.13013

Cree BAC, Cutter G, Wolinsky JS, Freedman MS, Comi G, Giovannoni G, Teams SPII (2020) Safety and efficacy of MD1003 (high-dose biotin) in patients with progressive multiple sclerosis (SPI2): a randomised, double-blind, placebo-controlled, phase 3 trial. Lancet Neurol 19(12):988–997. https://doi.org/10.1016/S1474-4422(20)30347-1

Eriksen AB, Berge T, Gustavsen MW, Leikfoss IS, Bos SD, Spurkland A, Blomhoff HK (2015) Retinoic acid enhances the levels of IL-10 in TLR-stimulated B cells from patients with relapsing-remitting multiple sclerosis. J Neuroimmunol 278:11–18. https://doi.org/10.1016/j.jneuroim.2014.11.019

Evans E, Piccio L, Cross AH (2018) Use of vitamins and dietary supplements by patients with multiple sclerosis: a review. JAMA Neurol 75(8):1013–1021. https://doi.org/10.1001/jamaneurol.2018.0611

Fanara S, Aprile M, Iacono S, Schirò G, Bianchi A, Brighina F, Salemi G (2021) The role of nutritional lifestyle and physical activity in multiple sclerosis pathogenesis and management: a narrative review. Nutrients 13(11):3774

Ferland G (2012) Vitamin K and the nervous system: an overview of its actions. Adv Nutr 3(2):204–212. https://doi.org/10.3945/an.111.001784

Filippi M, Preziosa P, Banwell BL, Barkhof F, Ciccarelli O, De Stefano N, Rocca MA (2019) Assessment of lesions on magnetic resonance imaging in multiple sclerosis: practical guidelines. Brain 142(7):1858–1875. https://doi.org/10.1093/brain/awz144

Gn B, Mm O, An B, IuB MN, Popova NF (2011) Possibilities of treatment of multiple sclerosis exacerbations without corticosteroids: a role of metabolic and antioxidant therapy. Zhurnal nevrologii i psikhiatrii imeni S.S Korsakova 111(2):44–48

Jafarirad S, Siassi F, Harirchian MH, Amani R, Bitarafan S, Saboor-Yaraghi A (2013) The effect of vitamin a supplementation on biochemical parameters in multiple sclerosis patients. Iran Red Crescent Med J 15(3):194–198. https://doi.org/10.5812/ircmj.3480

Ji Z, Fan Z, Zhang Y, Yu R, Yang H, Zhou C, Ke Z-J (2014) Thiamine deficiency promotes T cell infiltration in experimental autoimmune encephalomyelitis: the involvement of CCL2. J Immunol 193(5):2157–2167

Khosravi-Largani M, Pourvali-Talatappeh P, Rousta AM, Karimi-Kivi M, Noroozi E, Mahjoob A, Tavakoli-Yaraki M (2018) A review on potential roles of vitamins in incidence, progression, and improvement of multiple sclerosis. eNeurological Sci 10:37–44. https://doi.org/10.1016/j.ensci.2018.01.007

Løken-Amsrud KI, Myhr K-M, Bakke SJ, Beiske AG, Bjerve KS, Bjørnarå BT, Holmøy T (2013) Retinol levels are associated with magnetic resonance imaging outcomes in multiple sclerosis. Mult Scler J 19(4):451–457. https://doi.org/10.1177/1352458512457843

Maillart E, Mochel F, Acquaviva C, Maisonobe T, Stankoff B (2019) Severe transient myopathy in a patient with progressive multiple sclerosis and high-dose biotin. Neurology 92(22):1060–1062. https://doi.org/10.1212/wnl.0000000000007576

Marios CP, George NL, Evangelia EN, Ioannis SP (2013) A novel oral nutraceutical formula of omega-3 and omega-6 fatty acids with vitamins (PLP10) in relapsing remitting multiple sclerosis: a randomised, double-blind, placebo-controlled proof-of-concept clinical trial. BMJ Open 3(4):e002170. https://doi.org/10.1136/bmjopen-2012-002170

Mousavi Nasl-Khameneh A, Mirshafiey A, Naser Moghadasi A, Chahardoli R, Mahmoudi M, Parastouei K, Saboor-Yaraghi AA (2018) Combination treatment of docosahexaenoic acid (DHA) and all-trans-retinoic acid (ATRA) inhibit IL-17 and RORgammat gene expression in PBMCs of patients with relapsing-remitting multiple sclerosis. Neurol Res 40(1):11–17. https://doi.org/10.1080/01616412.2017.1382800

Offermanns S, Schwaninger M (2015) Nutritional or pharmacological activation of HCA(2) ameliorates neuroinflammation. Trends Mol Med 21(4):245–255. https://doi.org/10.1016/j.molmed.2015.02.002

Orti JER, Cuerda-Ballester M, Drehmer E, Carrera-Julia S, Motos-Munoz M, Cunha-Perez C, Lopez-Rodriguez MM (2020) Vitamin B1 intake in multiple sclerosis patients and its impact on depression presence: a pilot study. Nutrients 12(9):2655. https://doi.org/10.3390/nu12092655

Ortí JEDR, Cuerda-Ballester M, Drehmer E, Carrera-Juliá S, Motos-Muñoz M, Cunha-Pérez C, López-Rodríguez MM (2020) Vitamin B1 intake in multiple sclerosis patients and its impact on depression presence: a pilot study. Nutrients 12(9):2655

Pahan K, Schmid M (2000) Activation of nuclear factor-kB in the spinal cord of experimental allergic encephalomyelitis. Neurosci Lett 287(1):17–20. https://doi.org/10.1016/S0304-3940(00)01167-8

Rahimi-Dehgolan S, Masoudi M, Rahimi-Dehgolan S, Azimi AR, Sahraian MA, Baghbanian SM, Naser Moghadasi A (2023) Effect of vitamin a on recovery from the acute phase of multiple sclerosis-related optic neuritis, double-blind, randomized, placebo-controlled trial [original article]. Cas J Intern Med 14(1):23–30. https://doi.org/10.22088/cjim.14.1.23

Riccio P, Rossano R (2018) Diet, gut microbiota, and vitamins D + a in multiple sclerosis. Neurotherapeutics 15(1):75–91. https://doi.org/10.1007/s13311-017-0581-4

Runia TF, Hop WC, de Rijke YB, Hintzen RQ (2014) Vitamin a is not associated with exacerbations in multiple sclerosis. Mult Scler Relat Disord 3(1):34–39. https://doi.org/10.1016/j.msard.2013.06.011

Sainaghi PP, Collimedaglia L, Alciato F, Molinari R, Sola D, Ranza E, Avanzi GC (2013) Growth arrest specific gene 6 protein concentration in cerebrospinal fluid correlates with relapse severity in multiple sclerosis. Mediat Inflamm 2013:406483. https://doi.org/10.1155/2013/406483

Sama B, Aliakbar S-Y, Mohammad-Ali S, Danesh S, Shahriar N, Mansoureh T, Mohammad-Hossein H (2016) Effect of vitamin a supplementation on fatigue and depression in multiple sclerosis patients: a double-blind placebo-controlled clinical trial. Iran J Allergy Asthma Immunol 15(1):13–19

Sedel F, Papeix C, Bellanger A, Touitou V, Lebrun-Frenay C, Galanaud D, Tourbah A (2015) High doses of biotin in chronic progressive multiple sclerosis: a pilot study. Mult Scler Relat Disord 4(2):159–169. https://doi.org/10.1016/j.msard.2015.01.005

Stoiloudis P, Kesidou E, Bakirtzis C, Sintila SA, Konstantinidou N, Boziki M, Grigoriadis N (2022) The role of diet and interventions on multiple sclerosis: a review. Nutrients 14(6):1150. https://doi.org/10.3390/nu14061150

Tourbah A, Lebrun-Frenay C, Edan G, Clanet M, Papeix C, Vukusic S, Pelletier J (2016) MD1003 (high-dose biotin) for the treatment of progressive multiple sclerosis: a randomised, double-blind, placebo-controlled study. Mult Scler J 22(13):1719–1731. https://doi.org/10.1177/1352458516667568

Tryfonos C, Mantzorou M, Fotiou D, Vrizas M, Vadikolias K, Pavlidou E, Giaginis C (2019) Dietary supplements on controlling multiple sclerosis symptoms and relapses: current clinical evidence and future perspectives. Medicines 6(3):95

Chapter 8
The Potential Preventive and Therapeutic Role of Vitamin D in MS

Rayyah R. Alkhanjari, Maitha M. Alhajeri, Nadia Rabeh, Sara Aljoudi, Zakia Dimassi, and Hamdan Hamdan

Abstract This section provides a general background on vitamin D and investigates its role in preventing and treating multiple sclerosis (MS). Vitamin D, obtained from sunlight, a few dietary sources, and supplements, has shown promise in reducing the risk of MS development. It also possesses immunomodulatory properties that affect the immune cells and pathways involved in MS. Clinical trials suggest that vitamin D supplementation can reduce disease activity and enhance the quality of life in MS patients. However, further research is needed to determine optimal dosages and long-term benefits. Understanding vitamin D's potential in MS management offers hope for improving the lives of those with this complex condition.

Keywords Multiple Sclerosis (MS) · Vitamin D · Vitamin D receptor (VDR) · Vitamin D Deficiency · Autoimmune diseases

Abbreviations

BEYOND	Betaferon Efficacy Yielding Outcomes of a New Dose trial
BMI	Body mass index
CD36	Cluster of differentiation 36
CHOLINE trial	Cholecalciferol in relapsing-remitting MS: A randomized clinical trial

R. R. Alkhanjari · M. M. Alhajeri · N. Rabeh · S. Aljoudi · H. Hamdan (✉)
Department of Biological Sciences, College of Medicine and Health Sciences, Khalifa University, Abu Dhabi, United Arab Emirates
e-mail: hamdan.hamdan@ku.ac.ae

Z. Dimassi
Department of Medical Sciences, College of Medicine and Health Sciences, Khalifa University, Abu Dhabi, United Arab Emirates

© The Author(s), under exclusive license to Springer Nature Singapore Pte Ltd. 2024
H. Hamdan (ed.), *Exploring the Effects of Diet on the Development and Prognosis of Multiple Sclerosis (MS)*, Nutritional Neurosciences,
https://doi.org/10.1007/978-981-97-4673-6_8

COX	Cyclooxygenase
CTLA4	Cytotoxic T lymphocyte antigen
CYP2R1	Cytochrome P450 2R1
DBP	Vitamin D binding protein
IFN	Interferon
IL	Interleukin
iNOS	Inducible nitric oxide synthase
MR	Mendelian randomization
MS	Multiple sclerosis
NK	Natural killer
NPC1L1	Niemann-Pick C1-Like 1
PD1	Programmed death-1
PD-L1	PD1 ligand 1
PTH	Parathyroid hormone
RA	Rheumatoid Arthritis
RRMS	Relapse-remitting MS
RXR	Retinoid X receptor
SLE	Systemic lupus erythematosus
SNPs	Single-nucleotide polymorphisms
SOLAR	Supplementation of VigantOL® oil versus placebo as Add-on in patients with relapsing–remitting multiple sclerosis receiving Rebif® treatment
SOLARIUM	SOLAR ImmUne Modulating effects
SPF	sun protection factor
SR-B1	Scavenger receptor class B type 1
T1DM	Type 1 Diabetes Mellitus
TGF-β	Tissue growth factor-β
UV	Ultraviolet
vDBP	Vitamin D-binding proteins
VDDR	Vitamin D-dependent rickets
VDR	Vitamin D, Vitamin D receptor
VDRE	Vitamin D response elements
VIDAMS	Vitamin D to Ameliorate MS
vitamin D2	Ergocalciferol
vitamin D3	Cholecalciferol
vitamin D4	22-dihydroergocalciferol

Learning Objectives
- Identify the various sources of vitamin D, including sunlight, dietary sources, and supplements.
- Describe how vitamin D influences the development and management of MS through its immunomodulatory properties.

- Assess the findings of clinical trials regarding the effectiveness of vitamin D supplementation in reducing disease activity and improving the quality of life in MS patients.
- Discuss the challenges and current research gaps in determining the optimal dosages and long-term benefits of vitamin D supplementation for MS management.
- Examine the various genetic, environmental, and lifestyle factors that affect vitamin D levels and their impact on MS outcomes.

8.1 Introduction to Vitamin D

The term "vitamin" was first coined in 1912 by Casimir Funk, who derived it from the words "vita," referring to life, and "amine," referring to a nitrogenous chemical necessary for living (Piro et al. 2010). Through extensive animal models research, Funk and others recognized that no animal could flourish or survive when feeding solely on the four major macromolecules: carbohydrates, proteins, lipids, and nucleic acids. For instance, a study by Frederick Gowland Hopkins illustrated that rats fed only the four macromolecules had compromised growth. Yet, their growth was significantly improved by adding milk, indicating the presence of "accessory factors" of "astonishingly small amounts" in the milk necessary for development. This is only one study of many emphasizing the role of vitamins in living organisms' well-being. Further work elaborated on the association between vitamin deficiencies and diseases like scurvy, beriberi, rickets, pellagra, and xerophthalmia, demonstrating that illnesses can be caused by factors other than pathogens or toxins (Semba 2012).

Vitamin D is among the classical vitamins appreciated in the late nineteenth century when rickets were identified as a low-calcium bone disease. Clinicians tried supplementing the affected children with calcium and phosphate, but no improvement was witnessed. However, when they developed interventions utilizing direct sunlight and cod liver oil, they documented a significant number of recoveries, leading to the discovery of the "calcium-depositing vitamin," named later as vitamin D (Janoušek et al. 2022; Jones 2018). Vitamin D is a fat-soluble vitamin that belongs to the steroid compounds and exists in the diet in two primary forms: Ergocalciferol (vitamin D2) and cholecalciferol (vitamin D3). Additional forms are vitamin D1, a historical name for vitamin D2 and lumisterol combination, and 22-dihydroergocalciferol (vitamin D4) found in fungi (Janoušek et al. 2022). The supply of vitamin D can be of exogenous or endogenous origin. Exogenously, Vitamin D can be obtained from fatty fish and fish oil, with prominent examples like cod and tuna liver oil, salmon, herring, mackerel, rainbow trout, and tilapia flesh (Janoušek et al. 2022). Notably, wild fish, not farming fish, is recommended since the former has a higher vitamin content (Lu et al. 2007). In addition to fish, vitamin D3 can be found in eggs, specifically the yolk, explaining the significance of vitamin supplementation in poultry. Dairy products rich in fat are also considered a good source of vitamin D, including butter, whipping cream, and cheese, which contain both vitamin D3 and, to some extent D2 (Janoušek et al. 2022). Animal products are not the only dietary source of vitamin D. In fact, mushrooms are the primary dietary source of vitamin D2. Ultraviolet (UV)-B light exposure is crucial for the vitamin production from ergosterol. Since most farmed mushrooms are grown in the dark, exposing them to the sun for 15 minutes is necessary to ensure proper conversion of ergosterol to ergocalciferol (Cardwell et al. 2018; Janoušek et al. 2022).

Endogenously, our skin is designed to produce vitamin D3 when exposed to sunlight. Once the UV-B rays penetrate the epidermis, they produce the photolysis of 7-dehydrocholesterol found in the keratinocyte cell membrane, converting it into pre-vitamin D3. The formed product is intrinsically unstable and would immediately isomerize to vitamin D3. The formed vitamin D3 will eventually be absorbed into the circulation and distributed around the body via binding to plasma-binding proteins (Bikle 2014; Janoušek et al. 2022). It is worth mentioning that the epidermal synthesis of vitamin D3 contributes 90% of our body's vitamin replenishment (Janoušek et al. 2022; Reichrath 2006). However, the challenge resides in ensuring adequate sun exposure to maintain physiologic levels of vitamin D while minimizing the risk of skin cancer. An experimental study investigated whether we could achieve optimal vitamin D3 levels through cutaneous synthesis while applying sunscreen to protect against sunburns and possible inflammation, both of which contribute to skin cancer. It was found that using sunscreens with sun protection factor (SPF) 15 and high UV-A protection (actual SPF were 17 and 18) prevented the development of erythema while allowing a statistically significant increase in vitamin D3 (Young et al. 2019). In a review paper, Jörg Reichrath explained that exposing less than 18% of the body surface, which includes the face, arms, and hands, to a range of minimal erythema dose of one-third to one half (about 5 min for skin type-2 adult in Boston at noon in July) two or three times a week can be equivalent to ingesting about 250 mg (10,000 IU) of vitamin D. He emphasized the use of sunscreen for more prolonged exposures compared to the recommended amount (Reichrath 2006). Some evidence has linked vitamin D to skin cancer. A recent meta-analysis showcased a positive relationship between circulating 25-hydroxyvitamin D (25(OH)D) levels and the risk of melanoma and keratinocyte cancers, namely squamous and basal cell carcinoma, while also suspecting that sun exposure may be a possible confounder. On the other hand, dietary and supplemental vitamin D was only associated with basal cell carcinoma risk (Mahamat-Saleh et al. 2020). In conclusion, protected sun exposure is crucial, and with sufficient dietary intake, we can ensure an adequate level of vitamin D to support various physiological processes.

8.2 Vitamin D Absorption and Metabolism

Vitamin D is a fat-soluble vitamin that is assumed to follow the same steps of lipid digestion and absorption. The digestive process begins in the stomach, where the gastric acid and pepsin release the dietary non-hydroxylated vitamin D from the food matrix. Once in the duodenum, the pancreatic enzymes continue releasing the vitamin from the surrounding matrix. Then, it will be emulsified and solubilized into micelles with the help of bile salts to be absorbed through the lymphatic ducts. On the contrary, given the enhanced solubility of the hydroxylated forms of vitamin D in the blood, their absorption occurs through the portal system. Until this day, the exact intestinal location where vitamin D absorption occurs in humans is unknown. Based on animal studies, it is hypothesized that the jejunum and the ileum primarily mediate the uptake of the vitamin. Specific apical proteins, namely scavenger receptor class B type 1 (SR-B1), cluster of differentiation 36 (CD36), and Niemann-Pick C1-Like 1 (NPC1L1), are required to facilitate the absorption. Passive diffusion has been suggested as an additional absorptive mean at pharmaceutical concentrations. Once in the enterocytes, non-hydroxylated vitamin D is incorporated into the chylomicrons to be transported into the lymphatic system, bypassing the first-pass metabolism. Finally, the distribution of the vitamin and its metabolites hinges on plasma-binding proteins called vitamin D-binding proteins (vDBP) (Borel et al. 2015; Janoušek et al. 2022).

For both vitamin D2 and D3 to elicit their physiological response, they should first be converted to their active forms. The first step in the activation is the conversion to the 25-hydroxylated form (25(OH)D) via microsomal cytochrome P450 2R1 (CYP2R1) in the liver. The next step is the conversion of 25(OH)D to 1,25(OH)2D (also called calcitriol), which is the biologically active form. 1-a-hydroxylase, or CYP27B1, mainly in the kidney and other extrarenal tissues, catalyzes this step (Janoušek et al. 2022). Different factors have been shown to influence the activation process of vitamin D. For example, parathyroid hormone (PTH) stimulates CYP27B1, increasing the level of calcitriol, which in turn, through negative feedback, would block further synthesis by inhibiting PTH production and release (Chang and Lee 2019). Another well-recognized regulator is the fibroblast growth factor 23 (FGF23) and its cofactor α-klotho. They are mainly released during hyperphosphatemia, resulting in phosphate renal excretion, inhibition of CYP27B1, and activation of CYP24A1, which converts vitamin D and calcitriol into the inactive forms, 24,25(OH)2D and 1,24,25-(OH)3D, lowering the level of the active circulating vitamin D (Janoušek et al. 2022; Quarles 2012).

8.3 Physiological Role of Vitamin D

Vitamin D has multiple physiological effects, all of which are conveyed by its binding to the nuclear receptor, vitamin D receptor (VDR). Vitamin D binding to its receptor triggers the dimerization of VDR and the retinoid X receptor (RXR). The formed dimer, in turn, recruits other regulatory factors and binds to vitamin D response elements (VDRE), which are specific DNA sequences that can activate or repress genes (Bikle 2014; Janoušek et al. 2022). One of the vital functions of vitamin D is calcium homeostasis. It helps maintain normal serum calcium levels by promoting renal and intestinal calcium absorption. It also regulates bone mineralization and the function of osteoblasts and osteoclasts, preventing diseases such as rickets in children and osteomalacia and osteoporosis in adults (Bikle 2014; Chang and Lee 2019; Janoušek et al. 2022).

The immune functions of vitamin D are well documented. At the level of innate immunity, research has illustrated the capacity of calcitriol in potentiating the oxidative burst process by activating hydrogen peroxide synthesis in monocytes and inducing the production of antimicrobial agents of broad activity like cathelicidin by monocytes, epithelial cells, keratinocytes, lungs, and placental cells (Janoušek et al. 2022). The role of vitamin D in adaptive immunity was investigated by studying its influence on T cells. Experimental studies demonstrated that calcitriol can directly block T cell proliferation and specific pro-inflammatory cytokines production. Specifically, it can inhibit interleukin (IL)-2, IL-6, IL-12, IL-17, IL-23, and interferon (IFN)-γ. At the same time, it has the potential to stimulate the production of anti-inflammatory cytokines such as IL-10. In vitro research has also elucidated the role of vitamin D in inducing T regulatory cell development, which acts as immunomodulators (Bikle 2014; Cantorna et al. 2015; Janoušek et al. 2022). Moreover, vitamin D can target cytotoxic T cells and innate natural killer (NK) cells. A study by Chen et al. has shown that excluding the vitamin D effect by creating VDR knockout CD8+ T cells resulted in accelerated proliferation of cytotoxic T cells in the absence of antigen stimulation (Chen et al. 2014). Regarding NKs, vitamin D was shown to regulate cell survival in the thymus and control early cytokine synthesis of the cells (Cantorna et al. 2015). Overall, vitamin D plays a crucial role in balancing or curbing the immune response, mainly by down-regulating specific gene activity, as illustrated by a recent blood transcriptome analysis (Durrant et al. 2022).

8.4 Major Causes of Vitamin D Deficiency

The crucial role of vitamin D in different biological processes highlights the detrimental effect in cases of vitamin D deficiency. Specifically, as you continue reading this chapter, you will recognize the significant association between vitamin D and MS. The intriguing question remains: If we can endogenously synthesize it, what can cause vitamin D deficiency? Several factors have been implicated as potential causes, including environmental, dietary, and genetic. Among the main factors leading to vitamin D deficiency is reduced cutaneous synthesis. As stated above, cutaneous vitamin D synthesis depends on UV-B exposure; thus, inadequate exposure to sunlight resulting from small skin surface area exposed to the sun (i.e., whole-body clothing), living in high altitude or areas with extensive air pollution and cloud-shading, having dark skin, or applying sunscreen with high SPF (e.g., 30), can decrease the vitamin synthesis significantly (Chang and Lee 2019).

Regarding the exogenous vitamin D, any element that can interfere with the absorption stage can contribute to vitamin deficiency. Diseases characterized by impaired fat absorption, such as cystic fibrosis, obstructive jaundice, inflammatory bowel diseases, and pancreatic insufficiency, are associated with vitamin D deficiency since it follows the same fate as lipids (Borel et al. 2015; Janoušek et al. 2022). For example, a study has shown that cystic fibrosis patients absorbed less than a third to half of the dietary vitamin D2 compared to healthy individuals. Interestingly, the deficiency in this group of patients can be corrected by sun exposure and ingestion of the hydroxylated form of vitamin D (Borel et al. 2015). In addition to fat malabsorption, any process that interferes with the normal anatomy of the gastrointestinal tract can decrease oral vitamin D absorption. A study demonstrated a 25% reduction in the peak vitamin D3 after Roux-en-Y gastric bypass compared to the pre-operative values (Aarts et al. 2011).

It is important to note that deficiency in the active form of vitamin D can arise from abnormalities in the enzymes involved in its metabolism and activation. In fact, it is well-documented that people who suffer from chronic kidney disease have low calcitriol levels due to CYP27B1 impairment and suppression secondary to hyperphosphatemia. Liver diseases, including cirrhosis, can also diminish Vitamin D activation, as the 25-hydroxylation step will be affected. Moreover, some cases of hereditary vitamin D-dependent rickets (VDDR) are due to genetic mutations inducing alterations in vitamin D metabolism. For example, VDDR type 1 is due to a mutation in *CYP27B1* manifesting as low or no synthesis of 1,25(OH)2D renally. In VDDR type 3, there is a gain of function mutation in *CYP3A4*, which is transcribed and translated into an enzyme that inactivates vitamin D (Janoušek et al. 2022).

Pharmaceuticals constitute another critical factor influencing vitamin D levels and activity. For instance, medications such as orlistat, an inhibitor of gastric and pancreatic lipases, decrease vitamin D absorption by interfering with lipid digestion and absorption (Borel et al. 2015). Other drugs such as Phenytoin, carbamazepine, isoniazid, and rifampicin lower the levels of the active vitamin D by inducing CYP enzymes responsible for the catabolism and, consequently, the inactivation of the vitamin. By blocking the 1 α-hydroxylation, some medications can prevent vitamin D activation, namely the antifungals ketoconazole (Chang and Lee 2019; Janoušek et al. 2022). Those are only a few examples; many more can exist, but they are outside the scope of this book.

8.5 Autoimmune Diseases and Vitamin D Deficiency

Autoimmune diseases encompass a spectrum of complex disorders wherein the immune system erroneously targets the body's tissues and organs. They include a range of conditions such as Rheumatoid Arthritis (RA), lupus, type 1 diabetes, and multiple sclerosis (MS). In the subsequent section, we will delve into an in-depth discussion of vitamin D's role in the pathophysiology of MS.

Research shows that individuals who get more vitamin D through dietary intake or sunlight exposure have a lower risk of developing RA. These findings may be explained by the capacity of Vitamin D to control immune cells (Th1, Th17, Treg) involved in chronic joint inflammation (Sirbe et al. 2022). Additionally, a meta-analysis found that RA patients have either Vitamin D insufficiency or deficiency and documented a clear connection between lower vitamin D levels and the severity of RA symptoms, suggesting that vitamin D may be important in the risk of developing RA as well as the progression of the disease. Vitamin D supplementation is therefore a potential therapeutic adjunct in treating RA, pending more conclusive evidence. While there are some limitations in the existing studies, understanding the relationship between vitamin D and RA could advance our knowledge of autoimmune diseases and how to manage them (Lee and Bae 2016).

Next, exploring the connection between vitamin D and systemic lupus erythematosus (SLE), commonly known as Lupus, reveals a complex interplay. Lupus is an autoimmune disease affecting almost any organ system, but it mainly involves the skin, joints, kidneys, blood cells, and nervous system. Low vitamin D levels are frequently observed in individuals with Lupus, but it remains uncertain whether this deficiency is a consequence of the disease or contributes to its development (Islam et al. 2019; Touil et al. 2023). Vitamin D has been linked to a specific gene variant known as *CYP24A1*, which is associated with SLE risk. It was shown that having two copies of this gene variant and higher vitamin D levels lowers the risk of developing SLE in at-risk first and second-degree family members and distant relatives (Kendra et al. 2017). Conversely, individuals with vitamin D deficiency and two gene variants face a significantly increased risk of developing SLE. In some cases, the combination of deficiency and genetics accounts for 68.1% of SLE cases (Sirbe et al. 2022; Touil et al. 2023). A systematic review and meta-analysis by Islam et al. (2019) delved into vitamin D levels in Lupus patients compared to healthy individuals, revealing that Lupus patients generally have lower vitamin D levels in their blood. Various factors such as geographic location, latitude, body mass index (BMI), seasonal variations, medications, kidney disease, genetics, and modern lifestyle changes contribute to this deficiency. Interestingly, they also found that Lupus patients exhibit lower vitamin D levels even in regions with ample sunlight, possibly due to lifestyle factors, sun avoidance, or traditional veils. The factors listed are also expected to induce vitamin D deficiency in healthy individuals, but a factor that is conventionally seen in SLE patients is photosensitivity. Sunlight exposure can induce a flare-up in SLE patients, so they are advised to avoid sun exposure and apply sunscreen. Unfortunately, these protective measures make SLE patients more prone to developing vitamin D deficiency. Notably, the study showed that vitamin D supplementation can enhance vitamin D status in Lupus patients, emphasizing the importance of monitoring and addressing vitamin D deficiency in the management of SLE patients, especially when including glucocorticoids in the treatment regimen, which may contribute to vitamin D deficiency. Despite its limitations, this comprehensive analysis offered valuable insights into the global prevalence of vitamin D deficiency in individuals with SLE.

We now shift our attention to Type 1 Diabetes Mellitus (T1DM) and its potential connection to vitamin D. T1DM is characterized by the production of autoantibodies and autoimmune responses by Th1 and cytotoxic T cells, eventually leading to the destruction of insulin-producing pancreatic β cells (Sirbe et al. 2022). The exact relationship between T1DM and vitamin D remains unclear. While some studies suggested a possible association between low vitamin D levels and an increased risk of T1DM, others do not provide strong evidence for a direct cause-and-effect link (Manousaki et al. 2021; Sirbe et al. 2022). Interestingly, the authors of Mendelian randomization (MR) study argued that there might not be substantial evidence supporting a major impact of genetically determined lower vitamin D levels (25OHD) on the risk of developing T1DM. The authors explained that the results from observational studies could be influenced by environmental factors that need to be considered. They emphasized the need for further validation through large randomized controlled trials to confirm their findings (Manousaki et al. 2021). On a different note, a 24-month randomized controlled trial explored the effects of adding vitamin D to the conventional treatment regimen for T1DM. Surprisingly, the results showed an improvement in β-cell function as demonstrated by 2-hour C-peptide measurements and the reduced need for exogenous insulin among participants. This study was not without limitations. Due to the difficulty of recruiting participants, the authors could not include an arm to investigate the stand-alone effect of vitamin D treatment. Another limitation contributing to the reduced generalizability of the results is the population of individuals from uniquely Asian backgrounds (Yan et al. 2023). The relationship between T1DM and vitamin D is still an ongoing investigation. While some studies question the significance of genetically determined vitamin D levels as a predisposing risk factor for T1DM, others point to potential benefits in improving β-cell function through vitamin D supplementation (Yu et al. 2022). However, more research, including large-scale trials, is needed to provide definitive answers in this complex field.

In summary, the relationship between vitamin D and autoimmune diseases such as Rheumatoid Arthritis, Lupus, and Type 1 Diabetes Mellitus is intricate, and ongoing research should clarify its role and potential benefits as an adjunct therapy in managing these conditions. The following section provides an in-depth discussion of the relationship between vitamin D and MS.

8.5.1 Vitamin D and Its Effects on Inflammatory Markers in MS

A body of evidence has investigated the role of vitamin D in different immune processes to understand further its potential role in treating and preventing autoimmune diseases. Various studies have illustrated vitamin D's effect in modulating innate and adaptive immunity. It is shown that by binding to the VDR receptor on immune cells like lymphocytes, macrophages, and dendritic cells, vitamin D can regulate their growth and differentiation, which in turn can influence the incidence and prognosis of immune-related conditions (Dankers et al. 2016; Sirbe et al. 2022; Stoiloudis et al. 2022).

The general role of vitamin D as an immunomodulator has been extensively investigated. It was suspected that vitamin D could diminish the growth of helper T and B cells and the T cell response to autoantigens. It was also proven that vitamin D could increase the expression of different inhibitory markers, such as cytotoxic T lymphocyte antigen (CTLA4), programmed death-1 (PD1), and PD1 ligand 1

(PD-L1) on the helper T cells' surface, thus suppressing their proliferation (Sheikh et al. 2018). Interestingly, the effectiveness of vitamin D in repressing helper T cell proliferation may depend on sex, as some studies showed. Some animal and human in vitro research demonstrated that the vitamin was more effective in females than males, an observation that requires further investigation (Leffler et al. 2022). Moreover, vitamin D inhibits the production of pro-inflammatory mediators (Galoppin et al. 2022; Nielsen et al. 2017) by the binding of the activated VDR and RXR heterodimer to the vitamin D responsive elements in the promoter regions of B and T cells, therefore inhibiting the production of IFN-γ and IL-2 by CD4+ T cells (Nielsen et al. 2017; Sirbe et al. 2022; Yang et al. 2013).

Further, vitamin D can activate Th2 cells by increasing the production of IL-4, IL-5, IL-10, and tissue growth factor-β (TGF-β), shifting the immune response from a pro-inflammatory to an anti-inflammatory state (Galoppin et al. 2022; Yang et al. 2013). It has also been reported that VitD can enhance the activity of regulatory T cells and NK cells (Dankers et al. 2016; Yang et al. 2013). Regulatory T cells have been recognized to control and maintain immune homeostasis (Choi et al. 2016; Dankers et al. 2016). Vitamin D's exact role in enhancing those cells' activity can be attributed to the activation of the PD1: PD-L1 signaling pathway (Sheikh et al. 2018) (Fig. 8.1).

One of the prominent inflammatory markers in MS patients is the cytokine IL-6. A study by de la Rubia Orti et al. (de la Rubia Orti et al. 2021) conducted in Valencia on 39 patients with MS explored the relationship between vitamin D intake and MS progression by measuring plasma levels of IL-6. The study has found that patients with high levels of vitamin D intake exhibit low levels of IL-6, suggesting vitamin D exerts an anti-inflammatory effect. The study has also found that identifying the source of vitamin D intake is critical in determining the degree of inflammation. For instance, patients who obtain their vitamin D from diets that mainly include unsaturated fats are more likely to have less MS inflammatory-related responses. On the other hand, saturated fats and processed foods that contain vitamin D do not produce the same effects since these diets are associated with low vitamin D absorption. Despite the promising findings, this study is limited by its small sample size and its confinement to a specific population, making the results ungeneralizable (de la Rubia Orti et al. 2021).

Another study by Vandebergh et al. looked at the effects of serum 25(OH)D and BMI on the progression of MS using MR limited to European descents. The study has shown that lower BMI and genetically increased levels of serum 25(OH)D reduce the risk of developing MS, while low serum 25(OH)D is associated with worsening MS symptoms (Vandebergh et al. 2022). A similar MR study by Wang on a population of European origins illustrated comparable results, while also highlighting the adverse effects associated with the administration of elevated doses of vitamin D on the kidneys, CNS, and heart (Wang 2022). The recommended dosage for MS patients is yet to be determined. Thus far, the results of several therapeutic trials suggest that a daily oral or injection within the range of 3000 to 11,000 IU given for 3 months to 4 years will achieve serum levels within the advised range (Galoppin et al. 2022). Improved serum levels and signs of reduced disease activity

8 The Potential Preventive and Therapeutic Role of Vitamin D in MS

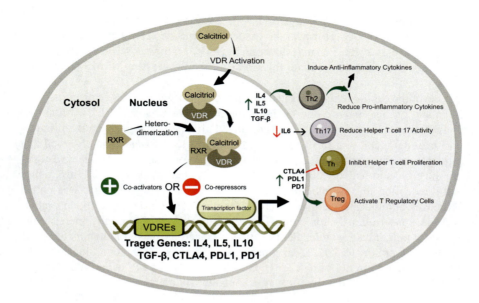

Fig. 8.1 The diagram illustrates the effects of vitamin D (the active form—calcitriol) in modulating inflammation by influencing the expression of different inflammatory markers. First, it crosses the cell and nuclear membranes via simple diffusion as it is a lipid-soluble vitamin. Once it is in the nucleus, it binds to its receptor, VDR. Then, the vitamin-receptor complex dimerizes with RXR. This complex can now bind to the VDREs on the DNA, modulating the expression of different genes. It can increase the expression of inhibitory signals such as CTLA4 and PD1: PDL-1, blocking the proliferation of helper T cells and inducing the activation of regulatory T cells. It can also stimulate the expression of different cytokines such as IL-4,5,10, and TGF-β, activating helper T cells type 2. This, in turn, shifts the immune response from a pro-inflammatory to an anti-inflammatory state. Further, by repressing the expression of cytokines like IL-6, vitamin D can decrease the activity of critical cells involved in MS pathogenesis, helper T cells type 17 (Galoppin et al. 2022; Korn and Hiltensperger 2021; Sheikh et al. 2018; Tesmer et al. 2008; Yang et al. 2013). *VDR* vitamin D receptor, *RXR* Retinoic X Receptor, *VDREs* vitamin D response elements, *CTLA4* cytotoxic T lymphocyte antigen, *PD1* programmed death-1, *PD-L1* PD1 ligand 1, *IL* interleukin

were observed when administering a daily dosage of 14,000 IU for a year (Sintzel et al. 2018; Sirbe et al. 2022).

A population-based case-control study investigated the association between neonatal levels of 25(OH)D, obtained from dried blood spot samples, and the risk of MS (controlling for multiple variables, including sex). They found that MS risk was the highest in individuals with vitamin D levels below 20.7 nmol/L at birth, while the lowest was in individuals above 48.9 nmol/L. The study also reported that with every 25 nmol/L increase in 25(OH)D, the odds of developing MS decreased by 30% (Nielsen et al. 2017). In a different study, Fitzgerald et al. explored the association between vitamin D levels and the progression of MS in relapse-remitting MS (RRMS) patients treated with INFβ-1b. The authors conducted a prospective cohort study based on the BEYOND (Betaferon Efficacy Yielding Outcomes of a New Dose trial) study and achieved a 31% reduction in the number of new active lesions among their study population with every 50 nmol/L increase in 25(OH) vitamin D ($P < 0.001$). However, no association was detected between vitamin D levels and relapse rate, change in brain volume, or the score on the Expanded Disability Status Scale—a scale for assessing the progression of MS (Fitzgerald et al. 2015).

One of the central regulators of CNS function is astrocytes; these cells protect the brain and further enhance and maintain the blood brain barrier (BBB). An abnormality in the morphology of these specialized glial cells can result in the progression of inflammation in MS patients. Interestingly, researchers have observed that astrocytes produce and interact with the active form of vitamin D. While investigating the effects of active vitamin D on MS EAE animal models and in vitro, a dramatic reduction in pro-inflammatory proteins, like TNF-α and IL-1β, was noted. Moreover, administering ergocalciferol (Vitamin D2 analog) to EAE models inhibited the transcription factor NF-κB, which is known to abnormally increase the release of inflammatory molecules that further exacerbate the inflammatory response in MS (Galoppin et al. 2022). An experimental study by Zhang et al. looked at the effects of paricalcitol (19-nor-1,25-(OH)2-vitamin D2) on MS using EAE models. The study results found a significant decline in the inflammatory components of the immune system, such as TNF-α, IL-1β, INF-γ, inducible nitric oxide synthase (iNOS), and cyclooxygenase (COX)-2 (Zhang et al. 2020). Furthermore, there was a marked decrease in the levels of F4/80 and CD3, which correlates with a reduction in the penetration of spinal cord tissues by macrophages and T cells, thereby leading to a reduced inflammatory response.

Although the studies mentioned above are examples of how vitamin D can influence the occurrence and progression of MS, randomized control trials reported little to no clinical significance. A double-blind, randomized control trial compared the effect of vitamin D3 to placebo on a population of RRMS patients taking IFN-β for 48 weeks (Smolders et al. 2019). The study was coined SOLAR (Supplementation of VigantOL® oil versus placebo as an add-on in patients with relapsing–remitting multiple sclerosis receiving Rebif® treatment). Initially, participants in the treatment group were given 6670 IU per day, which was then increased to 14,000 IU after 4 weeks. Although there were no safety issues, the trial's goals were not met as patients reaped no benefit from adding vitamin D to their regimen (Smolders et al. 2019). Another double-blind, parallel-group, placebo-controlled randomized clinical trial attempted to investigate the efficacy of vitamin D3 in MS treatment. The 96-week CHOLINE trial (Cholecalciferol in relapsing-remitting MS: A randomized clinical trial) administered 100,000 IU of vitamin D3 biweekly to the treatment group, while the control group was given a placebo. The goal of reducing patients' annual relapse rates was achieved in the treatment group, but the change was not statistically significant (Smolders et al. 2019).

The unremarkable results of the previous clinical trials inspired a study by Mimpen et al., which investigated the effect of numerous vitamin D metabolism-related polymorphisms on the serological response to high vitamin D supplementation in RRMS patients. A post hoc extended analysis of the SOLARIUM (SOLAR ImmUne Modulating effects) study, an extension of the SOLAR trial, examined four single-nucleotide polymorphisms (SNPs). Two SNPs related to vitamin D binding protein (DBP) (rs4588 and rs7041), a SNP related to the 1α-hydroxylase gene, *CYP27B1* (rs12368653), and a SNP related to the *CYP24A1* gene (rs2248359). It was reported that after 48 weeks of high vitamin D supplementation, carrying the DBP-related rs7041 allele lowered the serological response with a statistically significant reduction in the absolute increase of serum 25(OH)D. On the other hand, the CYP27B1-linked rs12368653 allele was associated with an elevated serological response in circulating 25(OH)D. These results show that vitamin D-related genetic environments are a potential confounder in supplementation studies. Equally important, it can affect the prognosis of MS, which needs extensive investigation to fully understand the possible therapeutic benefit of vitamin D for MS patients (Mimpen et al. 2021).

A recent report on the results of the VIDAMS (Vitamin D to Ameliorate MS) trial aimed to elucidate the effect of vitamin D3 doses on MS activity (Cassard et al. 2023). In this double-blind, randomized control trial, RRMS patients were enrolled and were allocated to one of two groups, a low-dose and high-dose group. The low-dose group was prescribed 600 IU daily, while the high-dose group was prescribed 5000 IU daily of vitamin D3 (Cassard et al. 2023). For 96 weeks, both groups consumed their supplements in addition to their daily regimen of glatiramer acetate, a traditional disease-modifying drug for MS treatment. The study found no statistically significant difference in the number of confirmed relapses in patients taking the low or high dose (Cassard et al. 2023). Therefore, the substantial vitamin D3 dosages used in the SOLAR, CHOLINE, and VIDAMS trials may not be advantageous in ameliorating the disease course.

The revelation of a genetic role in the interplay between vitamin D and MS serves as a catalyst for a more nuanced exploration of the variables influencing the response of MS patients to vitamin D (Mimpen et al. 2021). This discovery impels us to contemplate additional confounding or alternative factors that may contribute to the variability in outcomes, shedding light on the complex and multifaceted nature of the relationship between vitamin D supplementation and MS. Previously, we mentioned factors contributing to vitamin D deficiency, such as sun exposure, seasonal variations, geographical location, and BMI. These same factors could confound the results of studies investigating the association between vitamin D and MS activity. Future research should consider these factors when assessing the therapeutic potential of vitamin D. Studying the genetic and environmental influences on the interplay between vitamin D and MS activity can aid in tailoring personalized medicine plans for each patient. Other gaps in knowledge to address include the need to investigate the effect of vitamin D in different stages of MS. Most patients, approximately 85% of them, are of the RRMS phenotype, which lends them to an easy and accessible population to assess the effect of various therapeutic agents (Klineova and Lublin 2018). However, it is worth investigating the effect of vitamin D on other courses of the disease, such as clinically isolated syndrome, secondary progressive MS, progressive primary MS, and progressive relapsing MS.

8.6 Conclusion

The intricate interplay between vitamin D and the immune system, particularly in MS, has been extensively explored. While studies have highlighted the immunomodulatory effects of vitamin D and its potential to influence MS outcomes, the clinical significance of high dosages remains uncertain. Conflicting trial results and genetic factors influencing response emphasize the need for a personalized approach. Despite promising associations between vitamin D levels and reduced MS risk, optimal dosages and patient-specific considerations remain elusive. Future research should address gaps in knowledge, considering factors like sun exposure, geographical variations, and disease stages to refine our understanding and harness the therapeutic potential of vitamin D for MS patients.

8.7 Summary

This chapter explores the role of vitamin D in the prevention and treatment of MS. Vitamin D, derived from sunlight, dietary sources, and supplements, has shown promise in reducing the risk of developing MS and modulating the immune system. Clinical trials suggest that vitamin D supplementation can decrease disease activity and enhance the quality of life for MS patients. However, the optimal dosages and long-term benefits are still under investigation. The chapter highlights vitamin D's immunomodulatory effects, its role in regulating inflammatory markers, and the importance of understanding genetic and environmental factors influencing its efficacy. Future research is needed to refine dosage recommendations and develop personalized dietary strategies for individuals with MS, considering the potential impact of sun exposure, geographic location, and disease stages on vitamin D levels and MS outcomes.

References

Aarts E, van Groningen L, Horst R, Telting D, van Sorge A, Janssen I, de Boer H (2011) Vitamin D absorption: consequences of gastric bypass surgery. Eur J Endocrinol 164(5):827–832. https://doi.org/10.1530/eje-10-1126

Bikle DD (2014) Vitamin D metabolism, mechanism of action, and clinical applications. Chem Biol 21(3):319–329. https://doi.org/10.1016/j.chembiol.2013.12.016

Borel P, Caillaud D, Cano NJ (2015) Vitamin D bioavailability: state of the art. Crit Rev Food Sci Nutr 55(9):1193–1205. https://doi.org/10.1080/10408398.2012.688897

Cantorna MT, Snyder L, Lin YD, Yang L (2015) Vitamin D and 1,25(OH)2D regulation of T cells. Nutrients 7(4):3011–3021. https://doi.org/10.3390/nu7043011

Cardwell G, Bornman JF, James AP, Black LJ (2018) A review of mushrooms as a potential source of dietary vitamin D. Nutrients 10(10):1498. https://doi.org/10.3390/nu10101498

Cassard SD, Fitzgerald KC, Qian P, Emrich SA, Azevedo CJ, Goodman AD, Sugar EA, Pelletier D, Waubant E, Mowry EM (2023) High-dose vitamin D_3 supplementation in relapsing-remitting multiple sclerosis: a randomised clinical trial. eClinicalMedicine 59:101957. https://doi.org/10.1016/j.eclinm.2023.101957

Chang SW, Lee HC (2019) Vitamin D and health—the missing vitamin in humans. Pediatr Neonatol 60(3):237–244. https://doi.org/10.1016/j.pedneo.2019.04.007

Chen J, Bruce D, Cantorna MT (2014) Vitamin D receptor expression controls proliferation of naïve CD8+ T cells and development of CD8 mediated gastrointestinal inflammation. BMC Immunol 15:6. https://doi.org/10.1186/1471-2172-15-6

Choi IY, Piccio L, Childress P, Bollman B, Ghosh A, Brandhorst S, Suarez J, Michalsen A, Cross AH, Morgan TE, Wei M, Paul F, Bock M, Longo VD (2016) A diet mimicking fasting promotes regeneration and reduces autoimmunity and multiple sclerosis symptoms. Cell Rep 15(10):2136–2146. https://doi.org/10.1016/j.celrep.2016.05.009

Dankers W, Colin EM, van Hamburg JP, Lubberts E (2016) Vitamin D in autoimmunity: molecular mechanisms and therapeutic potential. Front Immunol 7:697. https://doi.org/10.3389/fimmu.2016.00697

de la Rubia Orti JE, Garcia MF, Drehmer E, Navarro-Illana E, Casani-Cubel J, Proano B, Sanchis-Sanchis CE, Escriva JD (2021) Intake of vitamin D in patients with multiple sclerosis in the Valencian region and its possible relationship with the pathogenesis of the disease. Life (Basel) 11(12):1380. https://doi.org/10.3390/life11121380

Durrant LR, Bucca G, Hesketh A, Moller-Levet C, Tripkovic L, Wu H, Hart KH, Mathers JC, Elliott RM, Lanham-New SA, Smith CP (2022) Vitamins D(2) and D(3) have overlapping but different effects on the human immune system revealed through analysis of the blood transcriptome. Front Immunol 13:790444. https://doi.org/10.3389/fimmu.2022.790444

Fitzgerald KC, Munger KL, Kochert K, Arnason BG, Comi G, Cook S, Goodin DS, Filippi M, Hartung HP, Jeffery DR, O'Connor P, Suarez G, Sandbrink R, Kappos L, Pohl C, Ascherio A (2015) Association of Vitamin D Levels with Multiple Sclerosis Activity and Progression in patients receiving interferon Beta-1b. JAMA Neurol 72(12):1458–1465. https://doi.org/10.1001/jamaneurol.2015.2742

Galoppin M, Kari S, Soldati S, Pal A, Rival M, Engelhardt B, Astier A, Thouvenot E (2022) Full spectrum of vitamin D immunomodulation in multiple sclerosis: mechanisms and therapeutic implications. Brain Commun 4(4):fcac171. https://doi.org/10.1093/braincomms/fcac171

Islam MA, Khandker SS, Alam SS, Kotyla P, Hassan R (2019) Vitamin D status in patients with systemic lupus erythematosus (SLE): a systematic review and meta-analysis. Autoimmun Rev 18(11):102392. https://doi.org/10.1016/j.autrev.2019.102392

Janoušek J, Pilařová V, Macáková K, Nomura A, Veiga-Matos J, Silva DDD, Remião F, Saso L, Malá-Ládová K, Malý J, Nováková L, Mladěnka P (2022) Vitamin D: sources, physiological role, biokinetics, deficiency, therapeutic use, toxicity, and overview of analytical methods for detection of vitamin D and its metabolites. Crit Rev Clin Lab Sci 59(8):517–554. https://doi.org/10.1080/10408363.2022.2070595

Jones G (2018) The discovery and synthesis of the nutritional factor vitamin D. Int J Paleopathol 23:96–99. https://doi.org/10.1016/j.ijpp.2018.01.002

Kendra AY, Melissa EM, Joel MG, Diane LK, Timothy BN, Gary SG, Michael HW, Mariko LI, Jennifer K, Patrick MG, Kathy HS, Rufei L, Daniel JW, David RK, John BH, Judith AJ, Jill MN (2017) Combined role of vitamin D status and $CYP_{24}A1$ in the transition to systemic lupus erythematosus. Ann Rheum Dis 76(1):153. https://doi.org/10.1136/annrheumdis-2016-209157

Klineova S, Lublin FD (2018) Clinical course of multiple sclerosis. Cold Spring Harb Perspect Med 8(9):a028928

Korn T, Hiltensperger M (2021) Role of IL-6 in the commitment of T cell subsets. Cytokine 146:155654. https://doi.org/10.1016/j.cyto.2021.155654

Lee YH, Bae SC (2016) Vitamin D level in rheumatoid arthritis and its correlation with the disease activity: a meta-analysis. Clin Exp Rheumatol 34(5):827–833. https://www.ncbi.nlm.nih.gov/pubmed/27049238

Leffler J, Trend S, Gorman S, Hart PH (2022) Sex-specific environmental impacts on initiation and progression of multiple sclerosis. Front Neurol 13:835162. https://doi.org/10.3389/fneur.2022.835162

Lu Z, Chen TC, Zhang A, Persons KS, Kohn N, Berkowitz R, Martinello S, Holick MF (2007) An evaluation of the vitamin D3 content in fish: is the vitamin D content adequate to satisfy the dietary requirement for vitamin D? J Steroid Biochem Mol Biol 103(3–5):642–644. https://doi.org/10.1016/j.jsbmb.2006.12.010

Mahamat-Saleh Y, Aune D, Schlesinger S (2020) 25-Hydroxyvitamin D status, vitamin D intake, and skin cancer risk: a systematic review and dose-response meta-analysis of prospective studies. Sci Rep 10(1):13151. https://doi.org/10.1038/s41598-020-70078-y

Manousaki D, Harroud A, Mitchell RE, Ross S, Forgetta V, Timpson NJ, Smith GD, Polychronakos C, Richards JB (2021) Vitamin D levels and risk of type 1 diabetes: a Mendelian randomization study. PLoS Med 18(2):e1003536. https://doi.org/10.1371/journal.pmed.1003536

Mimpen M, Rolf L, Poelmans G, van den Ouweland J, Hupperts R, Damoiseaux J, Smolders J (2021) Vitamin D related genetic polymorphisms affect serological response to high-dose vitamin D supplementation in multiple sclerosis. PLoS One 16(12):e0261097. https://doi.org/10.1371/journal.pone.0261097

Nielsen NM, Munger KL, Koch-Henriksen N, Hougaard DM, Magyari M, Jorgensen KT, Lundqvist M, Simonsen J, Jess T, Cohen A, Stenager E, Ascherio A (2017) Neonatal vitamin D status and risk of multiple sclerosis: a population-based case-control study. Neurology 88(1):44–51. https://doi.org/10.1212/WNL.0000000000003454

Piro A, Tagarelli G, Lagonia P, Tagarelli A, Quattrone A (2010) Casimir funk: his discovery of the vitamins and their deficiency disorders. Ann Nutr Metab 57(2):85–88. https://doi.org/10.1159/000319165

Quarles LD (2012) Skeletal secretion of FGF-23 regulates phosphate and vitamin D metabolism. Nat Rev Endocrinol 8(5):276–286. https://doi.org/10.1038/nrendo.2011.218

Reichrath J (2006) The challenge resulting from positive and negative effects of sunlight: how much solar UV exposure is appropriate to balance between risks of vitamin D deficiency and skin cancer? Prog Biophys Mol Biol 92(1):9–16. https://doi.org/10.1016/j.pbiomolbio.2006.02.010

Semba RD (2012) The discovery of the vitamins. Int J Vitam Nutr Res 82(5):310–315. https://doi.org/10.1024/0300-9831/a000124

Sheikh V, Kasapoglu P, Zamani A, Basiri Z, Tahamoli-Roudsari A, Alahgholi-Hajibehzad M (2018) Vitamin D3 inhibits the proliferation of T helper cells, downregulate CD4(+) T cell cytokines and upregulate inhibitory markers. Hum Immunol 79(6):439–445. https://doi.org/10.1016/j.humimm.2018.03.001

Sintzel MB, Rametta M, Reder AT (2018) Vitamin D and multiple sclerosis: a comprehensive review. Neurol Ther 7(1):59–85. https://doi.org/10.1007/s40120-017-0086-4

Sirbe C, Rednic S, Grama A, Pop TL (2022) An update on the effects of vitamin D on the immune system and autoimmune diseases. Int J Mol Sci 23(17):9784. https://doi.org/10.3390/ijms23179784

Smolders J, Torkildsen Ø, Camu W, Holmøy T (2019) An update on vitamin D and disease activity in multiple sclerosis. CNS Drugs 33(12):1187–1199. https://doi.org/10.1007/s40263-019-00674-8

Stoiloudis P, Kesidou E, Bakirtzis C, Sintila SA, Konstantinidou N, Boziki M, Grigoriadis N (2022) The role of diet and interventions on multiple sclerosis: a review. Nutrients 14(6):1150. https://doi.org/10.3390/nu14061150

Tesmer LA, Lundy SK, Sarkar S, Fox DA (2008) Th17 cells in human disease. Immunol Rev 223:87–113. https://doi.org/10.1111/j.1600-065X.2008.00628.x

Touil H, Mounts K, De Jager PL (2023) Differential impact of environmental factors on systemic and localized autoimmunity. Front Immunol 14:1147447. https://doi.org/10.3389/fimmu.2023.1147447

Vandebergh M, Dubois B, Goris A (2022) Effects of vitamin D and body mass index on disease risk and relapse Hazard in multiple sclerosis: a Mendelian randomization study. Neurol Neuroimmunol Neuroinflamm 9(3):e1165. https://doi.org/10.1212/NXI.0000000000001165

Wang R (2022) Mendelian randomization study updates the effect of 25-hydroxyvitamin D levels on the risk of multiple sclerosis. J Transl Med 20(1):3. https://doi.org/10.1186/s12967-021-03205-6

Yan X, Li X, Liu B, Huang J, Xiang Y, Hu Y, Tang X, Zhang Z, Huang G, Xie Z, Zhou H, Liu Z, Wang X, Leslie RD, Zhou Z (2023) Combination therapy with saxagliptin and vitamin D for the preservation of beta-cell function in adult-onset type 1 diabetes: a multi-center, randomized, controlled trial. Signal Transduct Target Ther 8(1):158. https://doi.org/10.1038/s41392-023-01369-9

Yang CY, Leung PS, Adamopoulos IE, Gershwin ME (2013) The implication of vitamin D and autoimmunity: a comprehensive review. Clin Rev Allergy Immunol 45(2):217–226. https://doi.org/10.1007/s12016-013-8361-3

Young AR, Narbutt J, Harrison GI, Lawrence KP, Bell M, O'Connor C, Olsen P, Grys K, Baczynska KA, Rogowski-Tylman M, Wulf HC, Lesiak A, Philipsen PA (2019) Optimal sunscreen use, during a sun holiday with a very high ultraviolet index, allows vitamin D synthesis without sunburn. Br J Dermatol 181(5):1052–1062. https://doi.org/10.1111/bjd.17888

Yu J, Sharma P, Girgis CM, Gunton JE (2022) Vitamin D and Beta cells in type 1 diabetes: a systematic review. Int J Mol Sci 23(22):14434

Zhang D, Qiao L, Fu T (2020) Paricalcitol improves experimental autoimmune encephalomyelitis (EAE) by suppressing inflammation via NF-kappaB signaling. Biomed Pharmacother 125:109528. https://doi.org/10.1016/j.biopha.2019.109528

Chapter 9
Role of Dietary Supplements in Multiple Sclerosis

Haia M. R. Abdulsamad, Amna Baig, Sara Aljoudi, Nadia Rabeh, Zakia Dimassi, and Hamdan Hamdan

Abstract This chapter explores various dietary supplements and their effects on the progression of multiple sclerosis (MS) by analyzing their impacts on the disease. This exploration extends to caffeine, carnitine, Coenzyme Q10, lipoic acid, Omega-3, N-acetylcysteine, and melatonin, elucidating their mechanisms and potential benefits in MS. The optimal dietary regimen for individuals with MS frequently involves incorporating high-fiber diets and specific dietary supplements or vitamins. A high-fiber diet alters the gut bacterial populations and exerts a beneficial effect on the gut microbiome, which not only enhances cognitive function but also reinforces the gut bacteria's well-established role in mitigating inflammation. Furthermore, a number of over-the-counter dietary supplements, such as caffeine, Omega-3, and melatonin, exhibit the potential to improve MS status. Vitamins have been shown to play a significant role in MS, most notably vitamin D, which has a well-established clinical role in MS pathogenesis and modulation. Concurrent investigations into other water and fat-soluble vitamins are unfolding, revealing promising findings that suggest potential benefits in addressing inflammation, motor symptoms, and psychological aspects associated with MS. This chapter underscores the diverse array of dietary supplements that could positively impact the physiological and psychological aspects of MS. From caffeine's influence on BBB integrity to carnitine's potential in mitigating fatigue, each supplement offers a unique avenue for exploration.

H. M. R. Abdulsamad · A. Baig · S. Aljoudi · N. Rabeh · H. Hamdan (✉)
Department of Biological Sciences, College of Medicine and Health Sciences, Khalifa University, Abu Dhabi, United Arab Emirates
e-mail: hamdan.hamdan@ku.ac.ae

Z. Dimassi
Department of Medical Sciences, College of Medicine and Health Sciences, Khalifa University, Abu Dhabi, United Arab Emirates

© The Author(s), under exclusive license to Springer Nature Singapore Pte Ltd. 2024
H. Hamdan (ed.), *Exploring the Effects of Diet on the Development and Prognosis of Multiple Sclerosis (MS)*, Nutritional Neurosciences, https://doi.org/10.1007/978-981-97-4673-6_9

Keywords Multiple sclerosis (MS) · Dietary Supplements · Caffeine · Omega-3 · Melatonin

Abbreviations

6-SM	6-sulfatoxymelatonin
A1AR	Adenosine 1A receptors
ALC	Acetyl-L-carnitine
BBB	Blood-brain barrier
BDI	Beck depression inventory
CNS	Central nervous system
CoQ-10	Coenzyme Q10
DHA	Docosahexaenoic Acid
EAE	Experimental autoimmune encephalomyelitis
EDSS	Expanded Disability Status Scale
FSS	Fatigue severity scale
ICAM-1	Intercellular adhesion molecule–1
IFN-γ-β-α	Interferon-gamma-beta-alpha
IgG	Immunoglobulin G
IL17A	Interleukin 17A
LFA-1	Leukocyte function-associated antigen-1
MBP	Myelin basic protein
MDA	Malondialdehyde
MMP-9	Matrix metalloproteinase-9
MS	Multiple sclerosis
MTNR1A	Melatonin receptor 1A
MSIS	Multiple Sclerosis Impact Scale
NAC	N-acetylcysteine
NF-κB	Nuclear transcription factor-kappa B
NO	nitric oxide
Olig-2	Oligodendrocyte transcription-2
OPCs	Oligodendrocyte progenitor cells
PBMC	Peripheral blood mononuclear cells
PHYS	Physical
Plp	Poly lipoproteins
PPAR	Peroxisome proliferator-activated receptor
PSYCH	Psychological
PUFA	Polyunsaturated fatty acids
QOL	Quality of life
ROR-α	Retinoid-related orphan receptor alpha
RORγt	Retinoic acid-related orphan receptor γ t

ROS	Reactive oxygen species
RRMS	Relapsing-remitting MS
RXR	Retinoic X receptor
SOD	Superoxide dismutase
TAC	Antioxidant capacity
Th	T helper
TNF	Tumor necrosis factor
VCAM-1	Vascular cell adhesion molecule-1
VLA-4	Very late antigen 4
Ω-3	Omega-3

Learning Objectives
- Identify various dietary supplements and their potential benefits in managing MS.
- Explain the mechanisms through which supplements like caffeine, carnitine, Coenzyme Q10, lipoic acid, Omega-3, N-acetylcysteine, and melatonin affect MS progression and symptoms.
- Assess the findings from clinical trials and studies regarding the efficacy of different dietary supplements in improving MS symptoms and overall patient well-being.
- Discuss how these dietary supplements interact with existing pharmacological treatments and their potential as adjunct therapies.
- Examine the challenges in determining optimal dosages, potential side effects, and the need for expert supervision when incorporating dietary supplements into MS treatment regimens.

9.1 Introduction to Dietary Supplements

Dietary Supplements can be incorporated in optimizing the therapeutic interventions for individuals with MS, as shown in Table 9.1. This section illustrates the role of some dietary supplements in the context of MS, which can modulate inflammatory mediators and their gene expression, thus influencing immune system cells such as Th1 and Th17 (Fanara et al. 2021; Orti et al. 2020), depression, fatigue, and cognitive status.

Caffeine, extensively studied as a prominent constituent of coffee, has been shown to improve the integrity of the blood-brain barrier (BBB), thanks to its hydrophobic nature and ability to act on the brain's receptors contribute to enhancing the BBB. The BBB is a crucial separation between the CNS and systemic circulation, the disruption of which leads to cerebrovascular abnormalities that represent early hallmarks of the development of MS (Herden and Weissert 2020; Minagar and Alexander 2003). Additionally, ongoing research explores the impact of carnitine, known for its dual function as a neurotransmitter and as a shuttle for fatty acids into mitochondria, with a specific focus on alleviating the common symptom of fatigue (Tryfonos et al. 2019).

Table 9.1 Summary of dietary supplement potential effect on MS

Supplement	Summary of effects on MS
Caffeine	1. Blocks A1AR and reduces adenosine transmission in the brain
	2. Improves the integrity of the BBB
	3. Upregulates A1AR and decreases pro-inflammatory cytokines such as TNF-α
	4. Improves fatigue, mental capacity[a], tiredness, and headaches
L-carnitine	1. Improves fatigue
	2. Improves motor coordination and balance
	3. Reduces oxidative stress in the CNS (reducing free radicals)
CoQ-10	1. Decreases serum amount of IL-6, TNF-α, MMP-9
	2. Lowers the levels of depression and fatigue
	3. Increases expression of MBP gene and Olig-1
	4. Increases levels of SOD and TAC
	5. Decreases in fatigue presentation and depression
Lipoic acid	1. Maintains the integrity of the BBB
	2. Limits the expression of the LFA-1, VLA-4, VCAM-1, ICAM-1, and MMP-9
	3. Reduces oxidative stress
	4. Decreases brain atrophy and improves walking speed
Omega-3 (Ω-3)	1. Reduces IL17A and RORγt gene expression
	2. Reduces PPAR-γ
	3. Reduces pro-inflammatory cells and cytokines
	4. Reduces differentiation of Th1 and Th17 cells
	5. Reduces inflammatory markers such as MMP-9, TNF-α, IL-1 and IL-6
N-acetylcysteine (NAC)	1. Improves cognition and attention
Melatonin	1. Prevents demyelination and inflammatory processes in MS
	2. Decreases numbers of Th17 cells in CNS, lymph nodes and spleen
	3. Increases IL-10 secreting CD4+ T cell and arrests in Th17 cell response

A1AR adenosine 1A receptors, *BBB* blood-brain barrier, *CNS* Central nervous system, *ICAM-1* intercellular adhesion molecule–1, *LFA-1* leukocyte function-associated antigen-1, *MMP-9* matrix metalloproteases 9, *MBP* myelin basic protein, *Olig-1* oligodendrocyte lineage gene 1, *RORγt* retinoic acid-related orphan receptor γ t, *SOD* superoxide dismutase, *TAC* total antioxidant capacity, *TNF-a* tumor necrosis factor-alpha, *VLA-4* very late antigen 4, *VCAM-1* vascular cell adhesion molecule-1
[a]In a patient with Expanded Disability Status Scale (EDSS) between 0–4

Furthermore, coenzyme Q10 (CoQ-10), recognized for its antioxidant properties, has demonstrated efficacy in reducing inflammatory markers such as TNF-α, IL-6, and MMP-9 in patients with MS (Evans et al. 2018; Sanoobar et al. 2015). Deficiency in Omega-3 (**Ω-3**) fatty acids, crucial for brain function, arguably increases the risk of myelin damage in MS patients (Tryfonos et al. 2019). N-acetylcysteine (NAC), a derivative of L-cysteine, emerges as a potential alternative resource for replenishing glutathione and addressing oxidative stress in MS (Banaclocha et al. 1997; Monti et al. 2020; Samuni et al. 2013; Viña et al. 1983). Lastly, melatonin, a key neurotransmitter in circadian rhythm, is explored for its potential role in preventing demyelination and inflammatory processes in MS (Tryfonos et al. 2019).

9.1.1 Caffeine

Caffeine, scientifically known as 1, 3, 7-trimethylxanthine, serves as a central nervous system (CNS) psychostimulant that belongs to the purine class. Its psychostimulant effect stems from its ability to reduce adenosine transmission in the brain, achieved through the blockade of adenosine receptors, particularly the adenosine 1A receptors (A1AR) (Fisone et al. 2004; Herden and Weissert 2018).

Through its action on A1R1 receptors leading to enhanced BBB integrity, caffeine could decrease the risk of developing MS in a dose-dependent manner (Evans et al. 2018). Upregulation of A1AR was shown to decrease pro-inflammatory cytokines such as tumor necrosis factor-alpha (TNF-a) (Herden and Weissert 2018), while The absence or decrease in the number of A1AR has been associated with severe MS, characterized by worsening demyelination, extensive axonal damage, and increased microglial activation (Tsutsui et al. 2004). The mechanism behind the upregulation of A1AR expression after long-term caffeine treatment is unknown. However, it has been suggested that it may be due to the repeated blockage of these receptors by caffeine (Shi et al. 1993).

Chen et al. (2008) studied the protective effects of caffeine on an intact BBB against a high-cholesterol diet designed to disrupt the BBB in the olfactory bulbs of rabbit models. The extent of BBB disruption was assessed by the extravasation of Immunoglobulin G (IgG) and fibrinogen, substances normally absent in the brain parenchyma. While increased extravasation of IgG and fibrinogen in the olfactory bulb was observed, rabbits with the same high-cholesterol diet and a daily intake of 3 mg of caffeine were protected against this extravasation. Overall, the results indicated that caffeine improves the integrity of the BBB and decreases the extravasation of autoimmune IgG, potentially limiting the pathogenesis of MS (Chen et al. 2008).

In terms of dosing, an estimated intake of 250–300 mg of caffeine, equivalent to 2–3 cups, has been shown to improve fatigue and mental capacity as well as alleviate tiredness and headaches in patients with an Expanded Disability Status Scale (EDSSsignificant cerebral dysfunction, in which the efficacy of caffeine in producing the desired effects may be diminished (Herden and Weissert 2020). Fatigue is a major disabling symptom in MS patients, leading to significant unemployment in this population, which makes caffeine a strong contender in improving the lifestyle of these individuals (Herden and Weissert 2020).

9.1.2 Carnitine (Acetyl-L-Carnitine)

Carnitine, recognized for its dual role as a neurotransmitter and as a shuttle for fatty acids into mitochondria for energy production, remains a subject of ongoing research in the context of MS (Tryfonos et al. 2019), with some evidence around the potential role of acetyl-L-carnitine (ALC) in alleviating fatigue, a common symptom among MS patients (Tomassini et al. 2004). An open-label study detected reduced carnitine levels in patients undergoing immunosuppressive treatment, and supplementation with 3–6 g of L-Carnitine for 3 months showed improvement in fatigue in 63% of patients, especially those treated with cyclophosphamide or interferon (Lebrun et al. 2006). In a cross-treatment clinical study addressing patients with MS-related fatigue, 70% of individuals receiving 1 g of ALC twice daily or amantadine 100 mg for 3 months exhibited a significant improvement in fatigue severity, as measured by the fatigue severity scale (FSS) (Tomassini et al. 2004). Exploring a cuprizone-MS-induced mouse model, carnitine improved motor coordination and balance, potentially attributed to increased expression of myelin-related genes such as Poly lipoproteins (*Plp*) and oligodendrocyte transcription-2 (*Olig-2*). ALC was also found to reduce oxidative stress in the CNS, contributing to a decline in free radicals. This reduction in oxidative stress creates a more favorable environment for oligodendrocyte progenitor cells (OPCs) to differentiate into mature oligodendrocytes capable of remyelination (Gharighnia et al. 2022).

Despite these promising findings, further investigations are warranted to conclusively establish the efficacy of ALC in improving fatigue associated with MS.

9.1.3 Coenzyme Q10 (CoQ-10) (Ubiquinone)

When administered to relapsing-remitting MS (RRMS) patients, Coenzyme Q10 (CoQ-10), known for its antioxidant properties, produced a decrease in the serum amount of IL-6, TNF-α, and matrix metalloproteinase–9 (MMP-9), in addition to reducing levels of depression and fatigue, but had no effect on IL-4 and TGF- β levels (Evans et al. 2018; Sanoobar et al. 2015). In a cuprizone-MS-induced mouse model, the administration of CoQ-10 was linked to improved myelination, indicated by an increase in the expression of the myelin basic protein (*MBP*) gene, a crucial component of the myelin sheath and the oligodendrocyte lineage gene 1 (*Olig-1*) (Khalilian et al. 2021). CoQ-10 also produced a reduction in oxidative stress, particularly in the corpus callosum (Khalilian et al. 2021). which was mediated by increased levels of superoxide dismutase (SOD) and antioxidant capacity (TAC), effectively inhibiting lipid peroxidation. SOD functions as a powerful antioxidant that protects neurons against the effects of reactive oxygen species (ROS). In contrast, TAC is a measurement used to assess the overall activity of antioxidants in the body (Khalilian et al. 2021; Wang et al. 2012).

As mentioned earlier, CoQ-10 supplementation was investigated for its influence on depression and fatigue in MS patients using the FSS and Beck depression inventory (BDI) questionnaire, respectively. Following a daily supplementation of 500 mg for 12 weeks, a significant decrease in both fatigue and depression was documented (Sanoobar et al. 2016).

9.1.4 Lipoic Acid

Lipoic acid is capable of maintaining the integrity of the BBB, thus preventing the infiltration of inflammatory cells. This protective action is achieved by limiting the expression of various molecules such as leukocyte function-associated antigen-1 (LFA-1), very late antigen 4 (VLA-4), vascular cell adhesion molecule-1 (VCAM-1), intercellular adhesion molecule–1 (ICAM-1), and MMP-9, thus protecting the brain endothelial cells. The reduction in the presentation of ICAM-1 and MMP-9 is attributed to a dose-dependent decrease in the activity of nuclear transcription factor-kappa B (NF-κB). This dose-dependent effect of lipoic acid on MMP-9 and ICAM-1 levels was evident in a pilot randomized controlled trial involving MS patients. Throughout the study, participants received varying doses of lipoic acid, ranging from 600 mg of lipoid acid taken twice daily, 1200 mg of lipoic acid once daily, or 1200 mg of lipoic acid taken twice daily, all over a span of 14 days. The study showed that higher doses correlated with elevated peak serum levels and increased effect on MMP-9 levels and ICAM-1 (Yadav et al. 2005).

Additional effects of lipoic acid include reduction of oxidative stress by neutralizing the effects of ROS and nitric oxide (NO) (Xie et al. 2022), decreased brain atrophy, and improved walking speed in mouse models. The underlying mechanism was not discussed (Evans et al. 2018).

9.1.5 Omega-3 (Ω-3)/Docosahexaenoic Acid (DHA)

Omega-3 (**Ω-3**), a crucial fatty acid, plays an important role in brain function, and its deficiency increases the risk of myelin damage. MS patients often exhibit low levels of eicosapentaenoic acid (EPA) and docosahexaenoic acid (DHA), both of which are components of **Ω-3** (Tryfonos et al. 2019).

Docosahexaenoic acid (DHA), an active component in omega-3, was found to reduce interleukin 17A (IL17A) and retinoic acid-related orphan receptor γ t (*RORγt*) gene expressions in a similar manner to vitamin A. This effect is attributed to DHA acting as a retinoic X receptor (RXR) agonist (Fig. 9.1). This proposed mechanism involves DHA entering peripheral blood mononuclear cells (PBMC), binding cytoplasmic RXR, and forming a heterodimer with the nuclear receptor PPAR-γ, which leads to its activation. The complex then enters the nucleus to suppress IL17A gene expression and causes the subsequent suppression of Th17

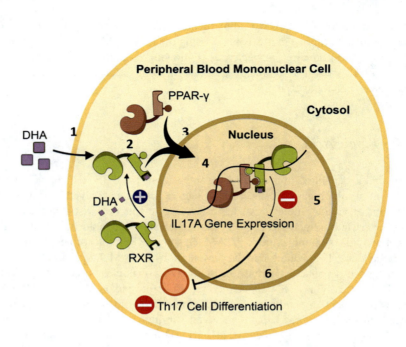

Fig. 9.1 A schematic diagram representing the proposed mechanism of action of DHA. DHA enters the PBMC (1), binds to RXR in the cytosol (2), and then forms a heterodimer with one of the nuclear receptors, transcription factor PPAR-γ (3). This dimerization leads to the activation of PPAR-γ (3). The complex enters the nucleus to (4) suppress IL17A gene expression (5), resulting in the suppression of Th17 differentiation (6) (Di Renzo et al. 2022; Mousavi Nasl-Khameneh et al. 2018). *Ω-3* Omega-3, *DHA* Docosahexaenoic acid, *RXR* retinoic X receptor, *PPAR-γ* Paeroxisome proliferator-activated receptor gamma, *IL17A* interleukin 17A, *Th17* T helper 17 cell

differentiation. The activity of RXR itself further inhibits IL17A gene expression, contributing to a decrease in disease activity (Di Renzo et al. 2022; Mousavi Nasl-Khameneh et al. 2018). Coadministration of DHA with vitamin A has a synergistic effect, reducing the pro-inflammatory cytokines IL17A and PPAR-γ, which are elevated in MS, thus reducing disease progression. These effects were studied in fifteen patients receiving interferon-gamma-beta-a (IFN-γ-β-α) for one year (Mousavi Nasl-Khameneh et al. 2018).

In experimental autoimmune encephalomyelitis (EAE), polyunsaturated fatty acids (PUFA) such as omega-3, along with its components EPA and DHA, generated a reduction in pro-inflammatory cells and cytokines. EPA reduced IFN-gamma and IL-17, while DHA reduced the differentiation of Th1 and Th17 cells. High doses of these components were found to decrease disability in RRMS by reducing inflammatory markers such as MMP-9, TNF-α, IL-1, and IL-6 (Fanara et al. 2021).

A low-fat diet Supplemented with Ω-3 for 1 year given in adjunct to pharmacological therapy showed moderate benefits in reducing the relapse rate in 31 RRMS patients (Weinstock-Guttman et al. 2005).

At the molecular level, Ω-3 has been shown to affect MMP-9 produced by immune cells. MMPs play a role in the pathogenesis of MS by disrupting the BBB and facilitating the migration of inflammatory cells to the CNS. Additionally, MPPs are involved in the cleavage of myelin basic protein, leading to demyelination. A three-month supplementation of Ω-3 decreased MMP-9 levels and immune cell secretion of MMP-9 by 58% (Shinto et al. 2009).

9.1.6 N-Acetyl-Cysteine

N-acetylcysteine (NAC) is a derivative of the natural amino acid L-cysteine and is readily available as an over-the-counter oral supplement in injectable forms. Functioning as a replenisher of glutathione (an antioxidant), NAC is recognized primarily as an antidote for acetaminophen overdose. Individuals with MS have documented low levels of glutathione levels in both the brain and periphery, increasing susceptibility to oxidative stress. Since oxidative stress takes part in MS pathogenesis, replenishing glutathione could have a role in MS treatment plans. However, direct supplementation of glutathione is challenging due to its pharmacokinetics features, making NAC a plausible alternative resource (Banaclocha et al. 1997; Monti et al. 2020; Samuni et al. 2013; Viña et al. 1983).

1. A study where NAC was supplemented alongside standard MS care for 2 months, using once-weekly injections and oral supplements six days a week, showed results (Monti et al. 2020). NAC led to increased glucose metabolism in certain brain areas, including the inferior frontal lobe, lateral and middle temporal lobes, and caudate, in parallel with improvements in attention and cognition, aligning with the regions responsible for these cognitive functions. Remarkably, no new lesions were observed, and existing lesions remained unchanged during the study. While this incidental finding hints at a potential role of NAC in MS treatment, further investigations are warranted to substantiate its impact. It is important to acknowledge that factors such as stress, depression, sleep, and fatigue might influence attention and cognition, which will need to be controlled in further investigations
2. Melatonin a major neurotransmitter in the circadian rhythm, is thought to play a role in preventing demyelination and inflammatory processes in MS (Tryfonos et al. 2019).

MS exhibits a degree of seasonality, with relapse incidence increasing during spring and summer and decreasing in fall and winter. Simultaneously, melatonin production increases in fall and winter during simulated darkness, suggesting a potential link between elevated levels of melatonin and reduced MS relapses in these seasons. Measurement of melatonin levels using 6-sulfatoxymelatonin (6-SM), a melatonin metabolite in urine, revealed interesting insights. Clinical symptoms of MS in an EAE-induced mice model were ameliorated when treated with melatonin (5 mg/kg) . This improvement was associated with a reduction in

Th17 cell number and frequency in the CNS, lymph nodes, and spleen, as well as a reduction in IL-17$^+$ IFN-γ$^+$ and IL-17$^+$ GM-CSF$^+$ CD4$^+$ T cells, all of which are implicated in EAE pathogenesis. The study also found an increase in IL-10 secreting CD4$^+$ T cells and an arrest in Th17 cell response (Farez et al. 2015).

Further investigations on 26 RRMS patients showed a correlation between higher levels of melatonin, lower IL17 and IL10 in peripheral CD4$^+$ T cells, and elevated IL10 expression, consistent with the findings in the mice model (Farez et al. 2015). Melatonin demonstrated an inhibitory effect on Th17 differentiation by decreasing the expression of IL-17 and IL-23 receptors on Th17 cells, independent of IFN-γ and IL-2 (Farez et al. 2015). Melatonin's interaction with nuclear retinoid-related orphan receptor alpha (ROR-α) and melatonin membrane receptor MTNR1A, expressed by Th17 cells, played a crucial role in modulating Th17 differentiation (Farez et al. 2015). Another mechanism through which melatonin inhibits Th17 cells exists, but its molecular intricacies are complex (Farez et al. 2015).

Melatonin also influenced IL-10 levels by activating MTNR1A and ROR-α on Tr1 cells, leading to an increase in IL-10 release(Farez et al. 2015). The quality of life (QOL) is low in MS patients due to poor sleep quality, depression, and fatigue. In a clinical study, 5 mg of melatonin was administered to 102 MS patients and 20 matched healthy subjects to investigate its effects on serum levels of malondialdehyde (MDA), SOD, and QOL. Findings showed variable effects of melatonin supplementation based on the patient's treatment regimen. The extent of the melatonin effect varied based on the patient's treatment. Overall, melatonin resulted in a reduction in the mean score MSIS (Multiple Sclerosis Impact Scale) -29-PHYS (Physical) and MSIS-29-PSYCH (Psychological) item-scores in the interferon beta-treated groups (Tryfonos et al. 2019).

In conclusion, it is essential to emphasize that the dietary supplements or vitamins mentioned in this book do not replace the need for pharmacological treatments and should be administered under expert supervision (Adamczyk-Sowa et al. 2014).

9.2 Conclusion

In conclusion, the exploration of dietary supplements, including caffeine, carnitine, Coenzyme Q10, lipoic acid, Omega-3, N-acetylcysteine, and melatonin, has revealed promising avenues for enhancing the management of MS. These supplements exhibit diverse physiological and psychological benefits impacting inflammation, fatigue, and cognitive function. While presenting potential adjuncts to MS treatment, it is imperative to acknowledge that they do not substitute pharmacological interventions. Expert supervision is essential for their judicious use in conjunction with established treatments. The ongoing quest for optimal dosages and regimens underscores the need for further research to solidify their role in the comprehensive care of individuals with MS.

9.3 Summary

This chapter delves into the impact of various dietary supplements on MS progression, focusing on their individual mechanisms and benefits. It covers supplements such as caffeine, carnitine, Coenzyme Q10, lipoic acid, Omega-3, N-acetylcysteine, and melatonin. High-fiber diets and specific vitamins, notably vitamin D, play significant roles in managing MS symptoms by enhancing gut microbiome health and exerting anti-inflammatory effects. Each supplement offers unique benefits, such as improving blood-brain barrier integrity, reducing fatigue, and alleviating cognitive dysfunction. While these supplements promise to be adjuncts to traditional pharmacological treatments, further research is needed to determine optimal dosages and ensure safety. Overall, the chapter highlights the diverse physiological and psychological benefits of dietary supplements in enhancing the quality of life for individuals with MS.

References

Adamczyk-Sowa M, Pierzchala K, Sowa P, Polaniak R, Kukla M, Hartel M (2014) Influence of melatonin supplementation on serum antioxidative properties and impact of the quality of life in multiple sclerosis patients. J Physiol Pharmacol 65(4):543–550

Banaclocha MMN, Hernández AI, Martınez N, Ferrándiz MAL (1997) N-acetylcysteine protects against age-related increase in oxidized proteins in mouse synaptic mitochondria. Brain Res 762(1):256–258. https://doi.org/10.1016/S0006-8993(97)00493-9

Chen X, Gawryluk JW, Wagener JF, Ghribi O, Geiger JD (2008) Caffeine blocks disruption of blood brain barrier in a rabbit model of Alzheimer's disease. J Neuroinflammation 5(1):12. https://doi.org/10.1186/1742-2094-5-12

Di Renzo L, De Lorenzo A, Fontanari M, Gualtieri P, Monsignore D, Schifano G, Alfano V, Marchetti M, Sierr. (2022) Immunonutrients involved in the regulation of the inflammatory and oxidative processes: implication for gamete competence. J Assist Reprod Genet 39(4):817–846. https://doi.org/10.1007/s10815-022-02472-6

Evans E, Piccio L, Cross AH (2018) Use of vitamins and dietary supplements by patients with multiple sclerosis: a review. JAMA Neurol 75(8):1013–1021. https://doi.org/10.1001/jamaneurol.2018.0611

Fanara S, Aprile M, Iacono S, Schirò G, Bianchi A, Brighina F, Dominguez LJ, Ragonese P, Salemi G (2021) The role of nutritional lifestyle and physical activity in multiple sclerosis pathogenesis and management: a narrative review. Nutrients 13(11):3774. https://www.mdpi.com/2072-6643/13/11/3774

Farez MF, Mascanfroni ID, Méndez-Huergo SP, Yeste A, Murugaiyan G, Garo LP, Balbuena Aguirre ME, Patel B, Ysrraelit MC, Zhu C, Kuchroo VK, Rabinovich GA, Quintana FJ, Correale J (2015) Melatonin contributes to the seasonality of multiple sclerosis relapses. Cell 162(6):1338–1352. https://doi.org/10.1016/j.cell.2015.08.025

Fisone G, Borgkvist A, Usiello A (2004) Caffeine as a psychomotor stimulant: mechanism of action. Cell Mol Life Sci 61(7):857–872. https://doi.org/10.1007/s00018-003-3269-3

Gharighnia S, Omidi A, Ragerdi Kashani I, Sepand MR, Pour Beiranvand S (2022) Ameliorative effects of acetyl-L-carnitine on corpus callosum and functional recovery in demyelinated mouse model. Int J Neurosci 134:409–419. https://doi.org/10.1080/00207454.2022.2107515

Herden L, Weissert R (2018) The impact of coffee and caffeine on multiple sclerosis compared to other neurodegenerative diseases. Front Nutr 5:133. https://doi.org/10.3389/fnut.2018.00133

Herden L, Weissert R (2020) The effect of coffee and caffeine consumption on patients with multiple sclerosis-related fatigue. Nutrients 12(8):2262. https://www.mdpi.com/2072-6643/12/8/2262

Khalilian B, Madadi S, Fattahi N, Abouhamzeh B (2021) Coenzyme Q10 enhances remyelination and regulate inflammation effects of cuprizone in corpus callosum of chronic model of multiple sclerosis. J Mol Histol 52(1):125–134. https://doi.org/10.1007/s10735-020-09929-x

Lebrun C, Alchaar H, Candito M, Bourg V, Chatel M (2006) Levocarnitine administration in multiple sclerosis patients with immunosuppressive therapy-induced fatigue. Mult Scler J 12(3):321–324. https://doi.org/10.1191/135248506ms1275oa

Minagar A, Alexander JS (2003) Blood-brain barrier disruption in multiple sclerosis. Mult Scler J 9(6):540–549. https://doi.org/10.1191/1352458503ms965oa

Monti DA, Zabrecky G, Leist TP, Wintering N, Bazzan AJ, Zhan T, Newberg AB (2020) N-Acetyl cysteine administration Is associated with increased cerebral glucose metabolism in patients with multiple sclerosis: an exploratory study [clinical trial]. Front Neurol 11:88. https://doi.org/10.3389/fneur.2020.00088

Mousavi Nasl-Khameneh A, Mirshafiey A, Naser Moghadasi A, Chahardoli R, Mahmoudi M, Parastouei K, Yekaninejad MS, Saboor-Yaraghi AA (2018) Combination treatment of docosahexaenoic acid (DHA) and all-trans-retinoic acid (ATRA) inhibit IL-17 and RORgammat gene expression in PBMCs of patients with relapsing-remitting multiple sclerosis. Neurol Res 40(1):11–17. https://doi.org/10.1080/01616412.2017.1382800

Orti JER, Cuerda-Ballester M, Drehmer E, Carrera-Julia S, Motos-Munoz M, Cunha-Perez C, Benlloch M, Lopez-Rodriguez MM (2020) Vitamin B1 intake in multiple sclerosis patients and its impact on depression presence: a pilot study. Nutrients 12(9):2655. https://doi.org/10.3390/nu12092655

Samuni Y, Goldstein S, Dean OM, Berk M (2013) The chemistry and biological activities of N-acetylcysteine. Biochim Biophys Acta 1830(8):4117–4129. https://doi.org/10.1016/j.bbagen.2013.04.016

Sanoobar M, Dehghan P, Khalili M, Azimi A, Seifar F (2016) Coenzyme Q10 as a treatment for fatigue and depression in multiple sclerosis patients: A double blind randomized clinical trial. Nutr Neurosci 19(3):138–143

Sanoobar M, Eghtesadi S, Azimi A, Khalili M, Khodadadi B, Jazayeri S, Gohari MR, Aryaeian N (2015) Coenzyme Q10 supplementation ameliorates inflammatory markers in patients with multiple sclerosis: a double blind, placebo, controlled randomized clinical trial. Nutr Neurosci 18(4):169–176. https://doi.org/10.1179/1476830513y.0000000106

Shi D, Nikodijević O, Jacobson KA, Daly JW (1993) Chronic caffeine alters the density of adenosine, adrenergic, cholinergic, GABA, and serotonin receptors and calcium channels in mouse brain. Cell Mol Neurobiol 13(3):247–261. https://doi.org/10.1007/BF00733753

Shinto L, Marracci G, Baldauf-Wagner S, Strehlow A, Yadav V, Stuber L, Bourdette D (2009) Omega-3 fatty acid supplementation decreases matrix metalloproteinase-9 production in relapsing-remitting multiple sclerosis. Prostaglandins Leukot Essent Fatty Acids 80(2-3):131–136. https://doi.org/10.1016/j.plefa.2008.12.001

Tomassini V, Pozzilli C, Onesti E, Pasqualetti P, Marinelli F, Pisani A, Fieschi C (2004) Comparison of the effects of acetyl L-carnitine and amantadine for the treatment of fatigue in multiple sclerosis: results of a pilot, randomised, double-blind, crossover trial. J Neurol Sci 218(1–2):103–108. https://doi.org/10.1016/j.jns.2003.11.005

Tryfonos C, Mantzorou M, Fotiou D, Vrizas M, Vadikolias K, Pavlidou E, Giaginis C (2019) Dietary supplements on controlling multiple sclerosis symptoms and relapses: current clinical evidence and future perspectives. Medicines 6(3):95. https://www.mdpi.com/2305-6320/6/3/95

Tsutsui S, Schnermann J, Noorbakhsh F, Henry S, Yong VW, Winston BW, Warren K, Power C (2004) A1 adenosine receptor upregulation and activation attenuates neuroinflammation and demyelination in a model of multiple sclerosis. J Neurosci 24(6):1521–1529. /10.1523/jneurosci.4271-03.2004

Viña J, Romero FJ, Saez GT, Pallardó FV (1983) Effects of cysteine and N-acetyl cysteine on GSH content of brain of adult rats. Experientia 39(2):164–165. https://doi.org/10.1007/BF01958877

Wang Y, Yang M, Lee SG, Davis CG, Koo SI, Chun OK (2012) Dietary total antioxidant capacity is associated with diet and plasma antioxidant status in healthy young adults. J Acad Nutr Diet 112(10):1626–1635. https://doi.org/10.1016/j.jand.2012.06.007

Weinstock-Guttman B, Baier M, Park Y, Feichter J, Lee-Kwen P, Gallagher E, Venkatraman J, Meksawan K, Deinehert S, Pendergast D, Awad AB, Ramanathan M, Munschauer F, Rudick R (2005) Low fat dietary intervention with omega-3 fatty acid supplementation in multiple sclerosis patients. Prostaglandins Leukot Essent Fatty Acids 73(5):397–404. https://doi.org/10.1016/j.plefa.2005.05.024

Xie H, Yang X, Cao Y, Long X, Shang H, Jia Z (2022) Role of lipoic acid in multiple sclerosis. CNS Neurosci Ther 28(3):319–331. https://doi.org/10.1111/CNS.13793

Yadav V, Marracci G, Lovera J, Woodward W, Bogardus K, Marquardt W, Shinto L, Morris C, Bourdette D (2005) Lipoic acid in multiple sclerosis: a pilot study. Mult Scler J 11(2):159–165. https://doi.org/10.1191/1352458505ms1143oa

Chapter 10
Plant-Based Extracts and Antioxidants: Implications on Multiple Sclerosis

Azhar Abdukadir, Rawdah Elbahrawi, Nadia Rabeh, Sara Aljoudi, Zakia Dimassi, and Hamdan Hamdan

Abstract Multiple sclerosis (MS) is a complex autoimmune disease characterized by inflammation, demyelination, and neurodegeneration, with oxidative stress playing a pivotal role in its pathogenesis. Recognizing the importance of mitigating oxidative damage, this abstract provides a succinct overview of the therapeutic potential of specific antioxidants—Ginkgo biloba, melatonin, curcumin, Edaravone, and N-acetyl cysteine—in the context of MS. Ginkgo biloba, a renowned herbal supplement, possesses robust antioxidant and anti-inflammatory properties. Preclinical phase studies suggest it can modulate immune responses, reduce neuroinflammation, and potentially impede MS progression. Melatonin, a potent antioxidant hormone, displays neuroprotective and immunomodulatory effects, positioning it as a promising adjunct therapy for MS. Curcumin, the active component of turmeric, demonstrates anti-inflammatory and antioxidant actions, offering potential benefits in reducing neuroinflammation and promoting remyelination. Edaravone, a free radical scavenger approved for neurodegenerative disorders, holds promise in alleviating MS-related oxidative stress. N-acetyl cysteine, a glutathione precursor, exhibits neuroprotective and anti-inflammatory properties, making it a potential therapeutic avenue for MS. This chapter will delve into the therapeutic potential of these antioxidants for MS management through a discussion of preclinical and clinical trials.

A. Abdukadir · R. Elbahrawi · N. Rabeh · S. Aljoudi · H. Hamdan (✉)
Department of Biological Sciences, College of Medicine and Health Sciences, Khalifa University, Abu Dhabi, United Arab Emirates
e-mail: hamdan.hamdan@ku.ac.ae

Z. Dimassi
Department of Medical Sciences, College of Medicine and Health Sciences, Khalifa University, Abu Dhabi, United Arab Emirates

© The Author(s), under exclusive license to Springer Nature Singapore Pte Ltd. 2024
H. Hamdan (ed.), *Exploring the Effects of Diet on the Development and Prognosis of Multiple Sclerosis (MS)*, Nutritional Neurosciences, https://doi.org/10.1007/978-981-97-4673-6_10

Keywords Multiple sclerosis (MS) · Plant-Based Extracts · Ginkgo biloba · Melatonin · Curcumin

Abbreviations

ALA	α-Lipoic acid
ALS	Amyotrophic lateral sclerosis
cAMP	Cyclic adenosine monophosphate
CB1	Cannabinoid 1
CB2	Cannabinoid 2
CBD	Cannabidiol
CLP	Cecal ligation and puncture
COX-2	Cyclooxygenase-2
DHA	Docosahexaenoic acid
EAE	Experimental autoimmune encephalomyelitis
EDSS	Expanded disability status scale
EGCG	Epigallocatechin-3-gallate
EPA	Eicosapentaenoic acid
G6PD	Glucose-6-phosphate dehydrogenase
GA	Glatiramer acetate
IFN	Interferon
IL	Interleukin
iNOS	Inducible nitric oxide synthase
MRI	Magnetic resonance imaging
MAO	Monoamine oxidase
MBP	Myelin basic protein
MPO	Myeloperoxidase
MS	Multiple sclerosis
NAC	N-acetyl cysteine
NOS	Nitro-oxidative stress
Nrf2	Nuclear factor-erythroid 2-related factor 2
OS	Oxidative stress
PET	Positron Emission Tomography
PKA	Protein kinase A
PKC	Protein kinase C
PRDX6	Peroxiredoxin 6
PUFAs	Polyunsaturated fatty acids
RG-I	Rhamnogalacturonan
RNS	Reactive nitrogen species
ROS	Reactive oxygen species
RRMS	Relapse remitting MS
SOD	Superoxide dismutase

SPMS	Secondary progressive MS
THC	Δ⁹-tetrahydrocannabinol
TNF	Tumor necrosis factor
Treg	Regulatory T

Learning Objectives
- Understand the therapeutic potential of specific antioxidants—Ginkgo biloba, melatonin, curcumin, Edaravone, and N-acetyl cysteine—in the management of MS.
- Describe the mechanisms through which these antioxidants exert their effects, including their roles in reducing oxidative stress, inflammation, and neurodegeneration in MS.
- Assess the findings from preclinical and clinical trials regarding the efficacy and safety of these antioxidants in improving MS outcomes.
- Understand the significance of oxidative stress in the pathogenesis of MS and how antioxidants can mitigate oxidative damage to improve patient outcomes.
- Identify the gaps in current research and the need for further clinical trials to establish the long-term benefits and optimal dosages of these antioxidants for MS patients.

10.1 Introduction

Multiple sclerosis (MS) is a chronic neurodegenerative condition, the hallmark features of which are demyelination and neuroinflammation, and that manifests as a broad spectrum of neurological symptoms. The root cause of MS remains unknown, with multiple factors contributing to its development, including oxidative stress (OS) and nitro-oxidative stress (NOS), particularly in the early phase of the disease (Siotto et al. 2019). OS or NOS can occur as a result of excessive generation of reactive oxygen species (ROS) and reactive nitrogen species (RNS) (e.g., superoxide), mitochondrial dysfunction, or poor antioxidant defense (Tobore 2021). ROS can be produced in various ways, including mitochondrial respiration, release from immune cells after an infection, and as oxidative enzyme by-products (Singh et al. 2019).

Through protein inactivation, OS leads to mitochondrial injury and dysfunction (Qi et al. 2006), which then results in reduced activity of mitochondrial energy metabolism and enzyme complexes seen in chronic MS active lesions (Lu et al. 2000). Oligodendrocytes and neurons are the most susceptible to being affected by OS, a feature attributed to their inherent poor antioxidant defense mechanisms. Injury to oligodendrocytes and neurons has been associated with axonal injury and demyelination. It has been observed that the attenuation of inciting factors of OS, such as ROS, facilitates the reversal of axonal degeneration and mitochondrial damage can be reversed. It is therefore worthwhile investigating MS therapies that mitigate and prevent OS, such as antioxidants (Tobore 2021), which act by increasing the function of enzymes promoting antioxidation (Adamczyk and Adamczyk-Sowa 2016). Research supports this theory as increased levels of OS markers were found in plasma isolated from MS patients (Ferretti et al. 2005). Consequently, it is believed that lowering OS is thought to aid in tissue regeneration and neuroprotection.

Antioxidants are natural or synthetically derived, enzymatic or non-enzymatic molecules (Hęś et al. 2019). They attenuate OS by reducing the concentration of oxidizable substrates. Within the body, naturally occurring antioxidative enzymes include NADH dehydrogenase 6, glucose-6-phosphate dehydrogenase (G6PD), superoxide dismutase (SOD), glutathione reductase, catalase, and glutathione peroxidase. ROS are directly eliminated by catalase, glutathione peroxidase, and SOD, while G6PD and glutathione reductase indirectly eliminate ROS by reducing levels of peroxide. Exogenously sourced antioxidants from the diet include compounds such as vitamins E, C, and A and beta-carotene. Polyphenolic compounds from leaves, seeds, flowers, spices, and tea also demonstrate antioxidative potential (Singh et al. 2019). Synthetic antioxidants, prepared to eliminate the cost and difficulty of isolating naturally occurring antioxidants, may have adverse effects such as allergic reactions and gastrointestinal disturbances (Lourenço et al. 2019).

10.2 Ginkgo Biloba

Ginkgo biloba, a natural herbal supplement, has been investigated for its possible therapeutic advantages in the treatment of MS, thanks to its antioxidant and neuroprotective properties, conferred by high concentrations of terpenoids and flavonoids. Ginkgo Biloba has garnered attention for its potential benefits in treating several other neurological conditions, including Alzheimer's disease and vascular dementia (Mojaverrostami et al. 2018).

Ginkgo biloba extract has been shown to have a neuroprotective effect against glutamate-induced excitotoxicity and apoptosis that could help preserve axonal integrity in MS (Zhang and Bhavnani 2006). Glutamate, an excitatory neurotransmitter, binds to glutamate receptors. It over-activates the receptor when upregulated, resulting in impaired cellular calcium homeostasis and free radicals, which will launch a cascade of excitotoxicity and apoptosis (Zhang and Bhavnani 2006).

Ginkgo biloba possesses anti-inflammatory properties and can influence the immune system by modulating the activity of T cells and other immune cells, potentially dampening autoimmune responses seen in MS. It acts by decreasing the production of pro-inflammatory cytokines tumor necrosis factor (TNF)-α interferon (IFN)-γ, (Zhang and Bhavnani 2006). and interleukin (IL)-15, which plays a vital role in the development of inflammatory protective immune responses for innate and adaptive immunity.

A systematic review examined Ginkgo Biloba's properties and how it could help improve cognitive performance in MS (Nabizadeh 2023). The review showed that excess levels of ROS initiated myelin phagocytosis and apoptosis in damaged nerve cells, leading to neurodegeneration and cognitive dysfunction in MS patients (Nabizadeh 2023). While Ginkgo Biloba did not help in increasing cognitive function in MS patients, it maintained their baseline level. Notably, however, evidence supporting the potential benefits of Ginkgo biloba in MS mainly comes from preclinical studies, as clinical evidence remains limited. Many in vitro and in vivo investigations have demonstrated the positive effects of ginkgo biloba supplementation on neurobehavioral outcomes, inflammatory indicators, and myelin repair. More human clinical trials are still needed to further support the use of Ginko Biloba in MS therapeutics (Lovera et al. 2012).

Generally, ginkgo biloba is regarded as safe when dosed properly, but it could still interact with several drugs as Ginkgo biloba has monoamine oxidase (MAO)-inhibitor properties, therefore synergistically interacting with MAO inhibitor drugs (Johnson et al. 2006). side effects include dry mouth, nausea, diarrhea and headache. Ginkgo biloba also possesses antiplatelet-activating factor properties, which could enhance clot formation and therefore should not be taken with anticoagulants (Sochocka et al. 2022). Overall, Ginkgo biloba is very safe for consumption; however, patients with MS should consult their healthcare providers before using Ginkgo biloba or any other supplements to avoid potential drug interactions or adverse effects.

10.3 Acer Truncatum Oil (Polyunsaturated Fatty Acids)

Acer truncatum oil, derived from the seeds of specific Acer species, is a rich source of polyunsaturated fatty acids (PUFAs). PUFAs are essential fatty acids that are crucial in maintaining cellular structure, regulating inflammation, and modulating the immune response (Farukhi et al. 2021). Acer truncatum oil contains a balanced ratio of omega-3 and omega-6 fatty acids, including eicosapentaenoic acid (EPA) and docosahexaenoic acid (DHA), which have well-documented anti-inflammatory and immunomodulatory properties (Zhang and Bhavnani 2006). Acer Tantrum Oil has shown promise in various neuroinflammatory conditions such as Alzheimer's disease and dementia and has potential therapeutic benefits in the management of MS.

Inflammation plays a central role in the pathogenesis of MS. Firstly, Acer Tantrum Oil, particularly EPA and DHA, has been shown to reduce the production of pro-inflammatory cytokines, such as IL-1β, TNF-α, and IL-6 (Kim et al. 2018), potentially dampening the neuroinflammatory processes, relieving symptoms and reducing disease progression in MS. Secondly, Acer Tantrum Oil protects against OS, a factor relevant to MS pathophysiology. Thirdly, Acer Tantrum Oil can modulate immune cell function by reducing the activation of T cells and promoting regulatory T (Treg) cell activity (Klingenberg et al. 2013). This immune-regulatory effect may help prevent the autoimmune response that underlies MS and may promote immune tolerance.

Thanks to its DHA content, Acer truncatum oil has demonstrated neuroprotective effects in neurodegenerating disease. DHA maintains neuronal membrane integrity and function, promotes the cell cycle exit, suppresses apoptosis, and promotes stem cell differentiation into neurons, improving neuronal survival and enhancing neuroplasticity (Kawakita et al. 2006). For example, the Zhang study used cuprizone, which is a copper chelator that destroys oligodendrocytes, thus causing demyelination. The study exposed mice to cuprizone for six weeks to induce demyelination. After the demyelination was confirmed by fluorescent immunohistochemistry and behavioral tests such as running wheels and footprint analyses, the cuprizone was withdrawn, and Acer truncatum oil was given for 2 weeks. The authors observed remyelination, increase in the maturity of oligodendrocyte cells, and upregulate the myelin basic protein (MBP) in the corpus callosum (Parlapani et al. 2009; Zhang and Bhavnani 2006). They concluded that Acer truncatum oil improves neuronal survival, enhances neuroplasticity, and has neuroprotective properties that could be a potential therapeutic benefit in managing MS.

In summary, experimental research has explored the potential benefits of Acer truncatum oil in experimental models of MS. Evidence has shown that Acer truncatum oil can reduce disease severity, decrease demyelination, and enhance remyelination (Xue et al. 2022). While promising, the therapeutic potential of Acer truncatum oil in MS warrants further investigation. Challenges include standardizing Acer truncatum oil formulations, optimizing dosing regimens, and conducting large-scale, well-designed clinical trials. Additionally, the long-term safety profile of Acer truncatum oil in MS patients needs thorough evaluation to suggest potential benefits in symptom management and quality of life improvements.

10.4 Ilex Paraguariensis St. Hilaire

Ilex paraguariensis St. Hilaire, also known as yerba mate, is a South American plant that is prepared as tea-like beverages and has stimulating effects due to its xanthine's-caffeine content (Cheminet et al. 2021; Filip et al. 2000). There are three important forms of consumption of Yerba mate, including chimarrão, an infusion of Yerba mate that is composed of 70% of the leaves and 30% of thin branches, tererê, a cold water extra of dried green leaves, and mate tea, hot water extract of warmed leaves (Camotti Bastos et al. 2018; Kungel et al. 2018). The Mate is prepared with two different infusion methods: one is prepared by adding boiled water to the dry plant, and the other is prepared by repeated close-to-boiled water to the dry plant. Both preparation methods are believed to allow an almost complete extraction of the water-soluble plant constituents (Filip et al. 2000). Yerba mate has important antioxidants, is vasodilatory, hypoglycemic, and simulates fat metabolism (Alkhatib and Atcheson 2017; Kungel et al. 2018). Yerba mate is rich in phenolic acids, caffeic and chlorogenic acids, and their derivatives, in addition to flavan-3-ols, (+)-catechin. Its therapeutic properties are attributed to the high caffeoyl derivates as caffeic acid, which exhibits antioxidant properties in biological systems by inhibiting a chemically initiated oxidation of liposomes (Bastos et al. 2007; Filip et al. 2000). Interestingly, a polysaccharide from leaves of yerba mate was purified, and its chemical structure was identified as a rhamnogalacturonan (RG-I), which exerts an anti-inflammatory effect by decreasing tissue expression of inducible nitric oxide synthase (iNOS) and cyclooxygenase-2 (COX-2) (Dartora et al. 2013).

The authors of this study aimed to evaluate the anti-inflammatory activity of RG-I against induced-polymicrobial sepsis in mice. The mice were randomly grouped into five clusters of 10 mice and underwent induced polymicrobial sepsis by cecal ligation and puncture (CLP), then every 12 h monitored for 7 days. The mice orally received either the RG-1 in 2, 7, or 10 mg/kg concentrations or water or were injected with dexamethasone subcutaneously. The control group did not undergo induced sepsis. Another set of mice were orally treated with RG-1 or dexamethasone 1 h before surgery. The mice that received the intervention 1 h before surgery were sacrificed 6 h post-operation. The lung and ileum tissues from the sacrificed mice were isolated to examine the myeloperoxidase (MPO) activity and tissue expression of iNOS and COX-2, respectively. It was observed that the lethality was markedly delayed in mice treated orally with the RG-1, with an overall survival rate of 20% for those who received the 2 mg/kg concentration, 40% for those who received the 7 mg/kg concentration, and 60% for those who received 10 mg/kg concentration ($p < 0.001$). The positive control group treated with dexamethasone showed a significant improvement in survival, with an overall survival rate of 16.7%. Ultimately, the results indicate that the RG-1 had a more significant percentage of survival rate compared to dexamethasone. MPO activity was of interest as it is a lysosomal enzyme of polymorphonuclear leukocytes that acts as a catalyst in the production of hypochlorous acid, a powerful oxidant. The results showed that CLP markedly increased the MPO levels in the lung tissue compared to the control group and that the RG-1 significantly prevented MPO rise in tissue, showing inhibition of 7.5%, 20.7%, and 38.4% at 2, 7, and 10 mg/kg respectively. In comparison, dexamethasone inhibited the activity in the lungs at only 33.7% ($p < 0.01$). Ultimately, the findings point out that RG-1 prevents the increase of MPO activity, thus indirectly reducing neutrophil recruitment to the lung and avoiding OS-induced tissue damage. iNOS and COX-2 levels in the ileum were determined by immunoblotting after different treatments. At 10 mg/kg, the polysaccharide decreased the levels of iNOS and COX-2 in the ileum by 29.5% and 32.3%, respectively. Dexamethasone reduced both iNOS and COX-2 expression by 43.2% and 38.7% ($p < 0.001$), respectively (Dartora et al. 2013). Although the study targeted experimentation in sepsis conditions, the properties exhibited by the polysaccharide show potential for inflammatory conditions.

A study adapted from Martins and Reissman 2007 was done to verify the total and hydrosoluble nutrient levels of yerba mate leaves and prepared Chimarrão in the safrinha harvest season and main harvest season. The study examined the total leaf contents like phosphorus [P], potassium [K], calcium [Ca], magnesium [Mg], sodium [Na], iron [Fe], manganese [Mn}, copper [Cu] and zinc [Zn], and further chemically analyzed the contents after acid digestion using a 3 mol L^{-1} hydrochloric acid [HCl] (Camotti Bastos et al. 2018). Analysis of boron content was obtained after dry ingestion and quantified with 0.45% azomethine-H (Camotti Bastos et al. 2018). The hydrosoluble content of the micronutrients listed above was examined by creating an infusion mixture of 3 g of milled leaves in 60 mL of deionized water heated to 80 °C that was allowed to be kept warm for 5 min and subsequently filtered through a blue-band paper filter. A 10-mL aliquot of the filtered extract was transferred to a crucible for evaporation, and the digestion was solubilized with 3 mol L^{-1} HCl and the concentrations of macro-and micronutrients were determined. They then stimulated Chimarrão using a mixture of 70% leaves and 30% fine branches that were then chemically analyzed using the same process described for the leaves. The results revealed that the leaves of yerba mate from the primary harvest season had the highest total nutrient content. In contrast, leaves from the secondary harvest season had higher contents of hydrosoluble nutrients. Much of the focus for this research was on the nutrient level, which is important in determining the overall health benefit of yerba mate. This research approach would benefit from investigating RG-1 and whether a suboptimal season or harvest affects function, availability, and long-term preservation for possible therapeutic use in MS.

10.5 Curcumin

The active compound of turmeric, curcumin, has antimicrobial, antioxidant, and anti-inflammatory properties (Bássoli et al. 2023). It has been used to treat cancers, gastrointestinal diseases, and neurodegenerative disorders such as MS (Atefi et al. 2021; Zoi et al. 2021). Curcumin can suppress the inflammatory response by inhibiting COX-2, an enzyme that produces prostaglandins that promote inflammation. Inflammatory cytokines such as IL-6 are also inhibited (Pulido-Moran et al. 2016). Its antioxidant capacity stems from curcumin's ability to directly eliminate free radicals as well as its propensity to stimulate the activity of endogenous antioxidants SOD and glutathione peroxidase (Atefi et al. 2021).

Preclinical research on mice models of MS treated with curcumin showcased the compound's immunomodulatory and antioxidant effects (ELBini-Dhouib et al. 2022). Before treatment with curcumin, the levels of endogenous antioxidants such as SOD were reduced after experimentally inducing MS in the mice. After treating the mice with curcumin, the levels of these endogenous antioxidants increased (ELBini-Dhouib et al. 2022). Interestingly, curcumin increased the degree of myelination to levels similar to that of mice in the negative control group. The anti-inflammatory properties of curcumin were evident after analyzing the mice's spinal cords stained with luxol fast blue and hematoxylin and eosin stains. Mice models of MS that did not receive curcumin showed increased demyelination and infiltration of inflammatory cells compared to their counterparts treated with curcumin (ELBini-Dhouib et al. 2022).

Given the promising results observed in vitro, the therapeutic potential of curcumin was investigated in clinical trials. Two randomized, double-blind, placebo-controlled trials using curcumin have demonstrated positive outcomes, with no adverse effects reported (Dolati et al. 2018; Petracca et al. 2021). The dosing of curcumin was kept at 80 mg/day or 80 mg/kg/day in either clinical trial (Dolati et al. 2018; Petracca et al. 2021). The dosage offering optimal results has not been examined in a clinical trial, nor can a comparison be drawn between clinical trials, as each trial measures different outcomes. Therefore, there is a need to understand the effect of curcumin dosage on MS activity. Although the trials do not comment on curcumin's antioxidant properties or report changes in OS markers, the trials demonstrated the compound's immunomodulatory effects. Another important finding was curcumin's ability to suppress Treg cells, thereby preventing auto-reactivity in immune cells (Dolati et al. 2019). A more recent randomized control trial in 2021 evaluated the efficacy of curcumin in patients with active relapsing MS (Petracca et al. 2021). The researchers measured clinical parameters such as disability and relapses, imaging parameters such as signs of inflammation on magnetic resonance imaging (MRI), and the tolerability and safety of curcumin as an adjuvant therapy to IFN-β. The high drop-out rate due to adverse effects, withdrawn consent, non-compliance, loss to follow-up, and alternative reasons lead to statistically insignificant differences in the measured parameters (Petracca et al. 2021). Due to inconclusive evidence, there remains a need for clinical trials assessing the utility, safety and tolerability of curcumin in the treatment of MS.

10.6 Epigallocatechin-3-Gallate (EGCG)

Epigallocatechin-3-gallate (EGCG) is a flavonoid and an active ingredient in green tea. Research shows that EGCG possesses immunomodulatory effects, which yield promising therapeutic results in treating chronic neurodegenerative diseases such as MS (Mähler et al. 2013). More specifically, EGCG can reduce populations of Th1 and Th17 cells and prevent the proliferation of autoreactive T cells and inflammatory cytokines. Along with its anti-inflammatory effects, EGCG can scavenge free radicals, protecting neurons from OS (Mähler et al. 2013). Its chemical structure contains phenols that can be readily oxidized to form quinone groups, endowing EGCG with antioxidant activity. Enhanced expression of endogenous antioxidants such as glutathione peroxidase and SOD was also found to be triggered by EGCG (Mähler et al. 2013).

Experimental research on mice models of MS demonstrates the beneficial effects of EGCG on the brain. In a sample of mice models of MS, administration of 50 mg/kg/day of EGCG for a duration of four weeks resulted in increased oligodendrocyte transcription factor 1, an essential mediator for myelin synthesis, and myelin proteolipid protein, crucial to form and maintain myelin's multilamellar structure (Miller et al. 2019). In vitro research found that EGCG with Glatiramer acetate (GA), an immunomodulator used in relapse remitting MS (RRMS), inhibited neuronal cell death, promoted axonal growth, delayed the onset of MS in animal models, improved disability status, and improved inflammation (Herges et al. 2011).

Several clinical trials are underway to investigate the safety and neuroprotective potential of EGCG in patients with MS (Plemel et al. 2015). The earliest randomized clinical trial aimed to investigate EGCG ability to reduce fatigue and muscle weakness in patients with RRMS (Mähler et al. 2015). It is hypothesized that EGCG improves metabolism and substrate processing, thereby reducing fatigue and muscle weakness. Patients were previously prescribed GA and then 600 mg/day of

EGCG was added to their regimen. EGCG was given for 12 weeks, and outcomes that were measured after the intervention included assessing energy expenditure, rises in fatty acid oxidation postprandially, and muscle work efficiency. The study did not explore the antioxidant effect of EGCG on patients, as measurements for markers of OS were not taken. However, the study found that EGCG improved muscle metabolism in men more so than women (Mähler et al. 2015). Another randomized placebo-controlled clinical trial recruited patients with progressive MS (Rust et al. 2021). For 36 months patients were started on 200 mg/day of EGCG, which was then increased to 400 mg/day at three months, then 600 mg/day at 6 months, 800 mg/day after 18 months, and 1200 mg/day after 30 months. The trial primarily assessed the rate of brain atrophy and safety through the analysis of adverse effects, medical and laboratory exams, as well as radiological and clinical parameters. Due to a small sample size and the majority of patients being in the non-relapsing stage of MS, no statistically significant differences in clinical or radiological parameters were seen in patients taking EGCG compared to placebo (Rust et al. 2021). Similar negative results were found in a randomized placebo-controlled clinical trial on patients with RRMS (Bellmann-Strobl et al. 2021)., where an 18-month intervention consisted of patients with RRMS taking up to 800 mg/day of EGCG as an adjuvant to GA therapy. The trial aimed to primarily assess the effectiveness and safety of EGCG by measuring radiological and clinical parameters (Bellmann-Strobl et al. 2021). Generally, EGCG was well tolerated and did not affect the body's immune response to GA. Mild adverse effects were observed in patients in the intervention and control groups, but none of the serious adverse events were attributed to the drug. Unfortunately, radiological evidence was inconclusive due to a small sample size and the potential patient-to-patient differences in plasma concentration of EGCG despite administering the same dosage (Bellmann-Strobl et al. 2021).

Positive outcomes of combined EGCG and coconut oil treatment on patients with MS were reported in a recent 2023 pilot study (Cuerda-Ballester et al. 2023). Patients were assigned to a placebo or intervention group receiving 800 mg of EGCG and 60 mL of coconut oil. Parameters such as gait, speed, and resistance were the primary outcomes measured in this study. Muscle strength was also measured by assessing serum levels of β-hydroxybutyrate and using a dynamometer. Gait speed and balance significantly improved in the intervention group, while gait resistance improved in both groups. Limitations of this trial were a lack of biochemical analysis for OS parameters and the small sample size. The study also lacked radiological or alternative methods of assessing nervous system activity.

10.7 α-Lipoic Acid

α-Lipoic acid (ALA) is a fatty acid exogenously sourced from foods such as broccoli, organ meat, and spinach, and endogenously synthesized in the liver (Plemel et al. 2015). Within the body, ALA has various roles such as the synthesis of coenzyme Q10 and glutathione, cofactors for mitochondrial enzymes like pyruvate dehydrogenase, and synthesis of vitamins C and E (Xie et al. 2022). It boasts robust anti-inflammatory and antioxidant potential. The anti-inflammatory response induced by ALA is made possible through the cyclic adenosine monophosphate (cAMP) and protein kinase A (PKA) pathway. Inflammatory interleukins such as IL-6 and IL-17 and T-cell proliferation were reduced upon ALA administration in patients with MS (Salinthone et al. 2010). Along with reducing pro-inflammatory responses, ALA can scavenge free radicals and chelate metals, thereby reducing OS (Plemel et al. 2015).

In preclinical studies on ALA in animal models of MS, there were improvements in demyelination and axonal damage. Interestingly, the extent of repair was dependent on the timing and mode of administration (Plemel et al. 2015). Axonal injury and demyelination were prevented when ALA was administered intra-peritoneally 7 days prior or immediately after inducing experimental autoimmune encephalomyelitis (EAE), an animal model used to study MS. Oral therapy was beneficial only when administered immediately after inducing EAE (Plemel et al. 2015). Research on animal models of MS also found that ALA could promote and maintain neuronal generation by increasing the number of mature oligodendrocytes (Xie et al. 2022).

Unlike other antioxidants, there are considerably more clinical trials using ALA for MS treatment. The earliest clinical trial was in 2005, which aimed to investigate the effect of oral ALA for tolerability, pharmacokinetics, and its effect on matrix metalloproteinase-9 and soluble intercellular adhesion molecule-1, which are proteins implicated in promoting the inflammatory response in the CNS (Yadav et al. 2005). Patients with MS were divided into one of four groups: a placebo group, daily 1200 mg of ALA group, a twice-daily dose of 1200 mg of ALA group, and a twice-daily dose of 600 mg of ALA group. The study lasted 2 weeks, and researchers found ALA to be well tolerated. Results showed that serum ALA varied among participants, and the highest serum ALA levels were achieved with the 1200 mg dose. A significant negative correlation between serum ALA and matrix metalloproteinase-9 was found. Similarly, a dose-response relationship between serum ALA levels and soluble intercellular adhesion molecule-1 was observed, with higher ALA levels leading to lower soluble intercellular adhesion molecule-1 levels (Yadav et al. 2005). Another randomized open-label trial aimed to assess and compare the pharmacokinetics of ALA in patients with MS and animal models of MS (Yadav et al. 2010). The trial found that patients receiving 1200 mg of ALA achieved a similar maximum concentration in the blood, Cmax, to that seen in mice receiving 50 mg/kg. The authors noted that the molecule's pharmacokinetics were influenced by the form in which ALA was administered. The capsulated formulation had a higher Cmax compared to tablets. However, these variations were attributed to the fact that patients took ALA with food instead of prior to eating. The results of the trial could not strongly confirm the clearance route ALA takes. It is speculated that at higher does, the non-renal metabolism of ALA predominates, and once this process is saturated, renal excretion of ALA ensues (Yadav et al. 2010).

Several other trials have focused on the pharmacokinetics of ALA. A randomized placebo-controlled trial administered a 1200 mg/day racemic mixture of ALA for 1 year to patients with secondary progressive MS (SPMS) (Bittner et al. 2017). After each dose, blood was drawn in consecutive intervals up to 240 min. The study found that participants' serum ALA levels remained consistently high throughout the year. Cmax did not change throughout the intervention, but variations between participants were noted. The variability in Cmax was attributed to factors such as age, drug-drug interactions, and aberrant gastrointestinal absorption (Bittner et al. 2017). It remains unknown whether these variations in Cmax impact the therapeutic potential of ALA. Another randomized crossover trial found that Cmax was the same in patients taking 600 mg of R-ALA or 1200 mg of R/S-ALA for 7–10 days (Cameron et al. 2020). However, the R-ALA form had better gastrointestinal tolerability and a lower rate of reported side effects.

A pilot study compared the two enantiomers of ALA in terms of their ability to upregulate the immunomodulatory compound, cAMP. ALA can exist as R-or S-ALA, depending on its optical rotation of polarized light (Salinthone et al. 2012). ALA can also exist as a racemic mixture (R/S-ALA). The study compared the racemic and R-ALA forms in patients with MS. After dividing patients into two groups and administering either form of ALA in a 1200 mg dose; blood was drawn at intervals up to 300 min after administration to quantify cAMP levels. Results showed that R/S-ALA took a shorter time to reach the maximum concentration in blood (Tmax). The levels of cAMP appeared to be higher upon R/S-ALA administration compared to R-ALA (Salinthone et al. 2012). The results of this pilot study are further supported by a clinical trial that measured the serum levels of cAMP after orally administering 1200 mg of ALA (Fiedler et al. 2018).

Clinical trials exploring the antioxidant potential of ALA are scarce, with only two clinical trials focusing on ALA and its ability to mitigate OS. A 2013 randomized clinical trial recruited patients with RRMS to investigate the effect of daily ALA consumption on OS (Khalili et al. 2014). Patients were given 1200 mg/day for 12 weeks, and blood samples were drawn 12 h after the last dose. The trial found that ALA improved total antioxidant capacity, but it did not enhance the activity of endogenous antioxidants such as SOD or glutathione peroxidase (Khalili et al. 2014). Another trial also confirmed the antioxidant capacity of ALA by measuring the formation of asymmetric dimethylarginine, which is formed in states of OS (Khalili et al. 2017). It was found that asymmetric dimethylarginine decreased with ALA administration, but there were no statistically significant differences in asymmetric dimethylarginine levels between patients taking ALA and the placebo group (Khalili et al. 2017).

Clinical trials on the efficacy of ALA in patients with SPMS show promising results. In a two-year double-blind, randomized clinical trial, patients with SPMS were given 1200 mg/day of ALA (Loy et al. 2018). outcome measures included improvements in gait, balance, and walking speed. It was found that walking ability was preserved with ALA consumption. More importantly, patients with an expanded disability status scale (EDSS) of less than 6 taking ALA were more likely to experience improvements in walking ability than the placebo group (Loy et al. 2018). Another clinical trial on patients with SPMS aimed to determine if ALA can slow brain atrophy at tolerable doses (Spain et al. 2017). Patients received 1200 mg/day of ALA or placebo for 2 years. The researchers primarily focused on measuring brain volume as well as rates of brain atrophy, safety, EDSS, and quality of life. The trial noted that the change in brain volume was significantly less in the ALA-treated group than in the placebo group. Compliance with the regimen was approximately 87%, but several adverse effects were noted, such as gastrointestinal pain, a case of glomerulonephritis, and a single case of renal failure (Spain et al. 2017). Nonetheless, the trial offers evidence of reduced brain atrophy with ALA consumption.

10.8 Cannabinoids

Cannabinoids are a collection of active compounds in plants such as *Cannabis sativa,* produced in animals or experimentally synthesized (Nouh et al. 2023). Of the many cannabinoids, the two most prominent ones are those found in the *Cannabis sativa* plant, cannabidiol (CBD) and Δ^9-tetrahydrocannabinol (THC). In animals, cannabinoids of different chemical structures are synthesized from arachidonic acid (Nouh et al. 2023). Specific G-protein coupled receptors cannabinoid 1 (CB1) and cannabinoid 2 (CB2), when bound by cannabinoids, can produce various physiological outcomes. Within the CNS, CB1 receptors are prevalent in neurons, and their role is to inhibit the release of neurotransmitters, while CB2 is more prevalent in immune cells with the role of regulating cytokine production and transport of immune cells through the CNS (Nouh et al. 2023). Like many agents discussed in this chapter, cannabinoids have anti-inflammatory and antioxidative properties.

There are direct and indirect ways in which cannabinoids act as antioxidative agents. Direct antioxidative effects occur through the inhibition of ROS formation or by enhancing endogenous antioxidant mechanisms. ROS generation is inhibited by inhibiting pro-oxidant enzymes xanthene oxidase and NADPH oxidase, chelating metal ions, and interrupting free radical chain reactions (Atalay et al. 2019). It supports endogenous antioxidant mechanisms by enhancing the activity of enzymes such as SOD, elevating levels of GSH, and preventing the depletion of microelements such as Zn. Indirect antioxidation methods are achieved by the interaction of cannabinoids with CB1 and CB2 receptors (Atalay et al. 2019). When activated, the CB1 receptor leads to the production of ROS and TNF-α, a pro-inflammatory cytokine. CB2 receptors antagonize CB1 receptors in that they prevent the production of ROS and TNF-α, thereby decreasing OS and inflammation. Cannabinoids, specifically CBD, are negative allosteric modulators at CB1 receptors but a weak agonist at CB2 receptors (Atalay et al. 2019). Preclinical studies have demonstrated the efficacy of cannabinoids in treating animal models of MS (Nouh et al. 2023). More specifically, cannabinoids have demonstrated the potential to reduce the immune response in animals with EAE. Treatment with cannabinoids has also demonstrated the potential to resolve symptoms of EAE in animal models (Nouh et al. 2023). Therefore, clinical studies have been pursued to assess the efficacy of cannabinoids in human subjects.

Clinical trials on the use of cannabinoids mainly utilized THC and CBD as interventions for patients with MS (Nouh et al. 2023). The trials completed thus far can be divided based on the drug used in the intervention. Earlier research on the use of THC mainly comprises case reports rather than clinical trials. A case report evaluating the efficacy of THC on eight patients with MS found that motor control was subjectively and objectively improved in two patients. In contrast, the others only experienced subjective improvements (Clifford 1983). An additional case report and pilot study further support the efficacy of THC in improving muscle spasticity (Brenneisen et al. 1996; Meinck et al. 1989). However, their results should be cautiously interpreted given their very small sample size and lack of generalizability. The earliest trial of the use of THC in humans dates back to 1981 (Petro and Ellenberger Jr 1981). The study aimed to evaluate muscle spasticity and strength in patients taking 5 or 10 mg of THC, orally, versus a placebo group. Patients in the intervention group reported improved walking ability and 10 mg of THC was able to reduce spasticity significantly. In patients with cerebellar disease, THC was contraindicated due to its relaxant properties (Petro and Ellenberger Jr 1981). A different placebo-controlled, double-blind crossover clinical trial assessed the efficacy of delta-9-THC in patients suffering from muscle spasticity due to MS (Ungerleider et al. 1988). Oral THC doses given to the patients incrementally increased from 2.5 to 15 mg for 5 days. It was found that doses greater than 7.5 mg improve patient-reported levels of muscle spasticity compared to patients taking a placebo (Ungerleider et al. 1988).

Synthetic oral formulations of THC have also been tested in patients with MS, with the results of official clinical trials still pending (Nouh et al. 2023). The two commonly used brand names are nabilone, containing about 1 mg of THC, and Dronabinol, which contains over 25 mg of THC. Two research letters report using nabilone in patients with MS, with detected improved muscle spasticity, reduced nocturia frequency, improvement in well-being, and analgesia (Nouh et al. 2023). Dronabinol was used in a randomized, placebo-controlled clinical trial consisting of stable patients with MS and muscle spasticity. The study aimed to demonstrate whether cannabinoids can improve MS symptoms and muscle spasms (Zajicek et al. 2003). Unlike the previous studies on patients with MS consuming THC, this trial did not observe improvements in muscle spasticity after cannabinoid consumption. However, the patient's perception of spasticity and mobility improved. Another placebo-controlled, double-blind, randomized crossover trial found that 10 mg of Dronabinol given daily for 3 weeks led to modest reductions in pain (Kristina et al. 2004).

The synergistic effects of CBD and THC in the management of patients with MS have been investigated in several clinical trials. These trials commonly use a drug under the trademarked name, Sativex, which consists of a mixture of THC and CBD with a 2.5:2.7 mg ratio (Nouh et al. 2023). The earliest placebo-controlled, double-blind, randomized crossover clinical trial assessed the safety and efficacy of THC and CBD, separately or combined, in alleviating neurogenic symptoms (Wade et al. 2003). Patients with MS were among the subjects recruited in this study and given the cannabinoids as a daily sublingual spray for 2 weeks. Although the study does not report the exact outcome of cannabinoid use in patients with MS, the authors generally note improvements in muscle spasticity, bladder control, and pain relief. The adverse events experienced were intoxication and brief hypotension upon initial dosing (Wade et al. 2003). A similar clinical trial but with an extended treatment period of 5 weeks also demonstrated improvements in pain relief upon Sativex consumption (Rog et al. 2005). This randomized, parallel-group, placebo-controlled clinical trial also noted reduced sleep disturbances. Apart from side effects such as dry mouth, dizziness, and somnolence, the treatment was well tolerated (Rog et al. 2005).

Muscle spasticity is a common symptom experienced by patients with MS (Collin et al. 2010). Clinical trials using Sativex in MS patients have successfully reduced muscle spasticity (Novotna et al. 2011). Patients were treated for 4 weeks in a double-blind, placebo-controlled, parallel-group trial, and those experiencing improvements in muscle spasticity continued the intervention for a total of 12 weeks. The trial found that muscle spasticity improved, and Sativex was generally well tolerated (Novotna et al. 2011). A clinical trial with a similar protocol compared the efficacy of Sativex as adjuvant therapy to first-line antispastic drugs (Markovà et al. 2019). The trial found significant improvements in muscle spasticity with Sativex as an adjuvant compared to placebo patients. Secondary outcomes such as EDSS, sleep disturbances, quality of life, and pain also significantly improved (Markovà et al. 2019).

The safety and efficacy of Sativex and cannabinoids have been investigated in clinical trials, which classify the drugs as generally safe for use (Patti et al. 2016; Zajicek et al. 2005). Some adverse events that have led to the discontinuation of the drugs include dizziness, fatigue, psychiatric symptoms, cognitive disturbances, allergic reactions, and oral and gastrointestinal discomfort (Patti et al. 2016). A major limitation of many clinical trials assessing the efficacy and safety of cannabinoids is a lack of long-term interventions. More recently, a clinical trial has shifted from the basic aim of previous trials assessing safety and efficacy to one in which the immunomodulatory mechanism is elucidated (Sorosina et al. 2018). Unfortunately, there remains a need for clinical trials investigating the efficacy of cannabinoids in the context of OS and antioxidant potential to achieve a better understanding of the antioxidative mechanism of cannabinoids.

10.9 Melatonin

Melatonin is a hormone secreted by the pineal gland and plays a vital role in regulating our circadian rhythm (Muñoz-Jurado et al. 2022). In addition to its role in sleep regulation, melatonin has also been characterized as an antioxidant. In preclinical studies on animal models of MS, it was found that melatonin triggers an increase in glutathione and endogenous antioxidative enzymes, such as SOD and glutathione peroxidase (Abo Taleb and Alghamdi 2020; Muñoz-Jurado et al. 2022). Melatonin is also thought to enhance the activity of nuclear factor-erythroid 2-related factor 2

(Nrf2), a regulator of antioxidative processes (Licht-Mayer et al. 2015). Originally, in the presence of OS, Nrf2 translocates from the cell membrane to the nucleus to upregulate anti-inflammatory and antioxidation genes. It is thought that this mechanism is not sufficient to prevent neuronal degeneration in the setting of MS (Licht-Mayer et al. 2015). Research has shown that melatonin supplementation is able to restore the antioxidative activity of Nrf2 (Fernández-Ortiz et al. 2020).

Currently, there are only a few clinical trials with published results on the use of melatonin in patients with MS. In a randomized control trial, the serum levels of IFN-β and vitamin B12 were measured after administration of 3 mg/day of melatonin for 24 weeks. It was found that serum levels of vitamin B12 and IFN-β increased in patients taking melatonin compared to the placebo group (Masoomeh et al. 2019; Yosefi-Fard et al. 2020). It is unclear exactly how and why melatonin led to these changes. Another randomized control trial focused on measuring inflammatory markers IL-1β and TNF-α in patients with MS taking 3 mg/day of melatonin for 24 weeks (Masoomeh et al. 2019). The study found a significant decrease in IL-1β but not in TNF-α (Masoomeh et al. 2019). A limitation of the current data on melatonin therapy in patients with MS is a lack of long-term interventions as well as a lack of trials focusing on measuring markers of oxidative stress after melatonin administration.

10.10 Edaravone

Edaravone is a pharmacological agent commonly used to treat amyotrophic lateral sclerosis (ALS) (Cha and Kim 2022). As a free radical scavenger, it works in neurodegenerative diseases, such as ALS, where OS is a contributing factor to their pathogenesis (Cha and Kim 2022). Given its neuroprotective effects under OS, this led researchers to pursue this drug in the setting of MS treatment, which is also induced by a state of OS. Preclinical research on Edaravone in EAE animal models found that the drug is neuroprotective as it decreases the expression of iNOS and lymphocytic infiltration of the spinal cord (Moriya et al. 2008). Another preclinical study on mice found that a combined treatment consisting of Edaravone and adipose-derived stem cells decreased demyelination and promoted the generation of oligodendrocytes (Bakhtiari et al. 2021). Edaravone's antioxidative mechanism of action in the context of MS has only been recently investigated. A study published in 2022 examined Edaravone's effect on ROS production in resting, protein kinase C (PKC) stimulated, and phagocytosis-stimulated granulocytes (Villar-Delfino et al. 2022). The of PKC and not other enzymes is based on the fact that PKC has been implicated in the production of ROS in neurodegenerative diseases (Villar-Delfino et al. 2022). The study recruited healthy controls and patients with MS, administered 1 μM of Edaravone, and withdrew participants' blood to determine the degree of ROS generation. Results of the study showed that Edaravone's antioxidant effect is primarily mediated by inhibiting the production of ROS in phagocytosis- and PKC-stimulated granulocytes (Villar-Delfino et al. 2022). Currently, there are no clinical trials investigating the therapeutic potential of Edaravone in patients with MS.

10.11 Ebselen

As a synthetic antioxidant, Ebselen plays a key role in lipid peroxidation by acting as a glutathione peroxidase mimetic. Promising research shows that even at low concentrations Ebselen has the potential to reduce the expression of inflammatory enzymes such as COX-2 and NADPH oxidase (Nakamura et al. 2002). A different study on mice with white matter lesions induced through chronic cerebral ischemia found that Ebselen could improve cognitive deficiencies and resolve the OS (Dong et al. 2023). The authors mention that the antioxidative mechanism was through upregulating the expression of antioxidative enzymes, promoting NRF2 activity, and promoting remyelination by elevating the levels of MBP (Dong et al. 2023). Despite the promising antioxidative effects of Ebselen, there remains a strong need for studies investigating its effect in animal models and patients with MS.

10.12 N-Acetyl Cysteine (NAC)

Glutathione is an endogenous antioxidant found in the body and is found to be decreased in patients with MS (Monti et al. 2020). Unfortunately, efforts to treat patients with MS using glutathione have been challenging due to its poor absorption and limited bioavailability. For this reason, NAC, known to increase glutathione levels, has been the focus of research efforts (Monti et al. 2020). In an exploratory study, researchers investigated the effect of NAC on cerebral glucose metabolism in patients with MS. Cerebral glucose metabolism was measured through Positron Emission Tomography (PET) scans of patients before and after NAC administration (Monti et al. 2020). The reason for measuring this metabolic parameter is due to its association with physiological brain function, which is often impaired in patients with MS (Roelcke et al. 1997). The researchers found that NAC improved cerebral glucose metabolism, leading to improved patient-reported cognitive capabilities (Monti et al. 2020). The promising results motivated researchers to pursue NAC in clinical trials in patients with MS. Currently, there is an ongoing trial aimed at assessing the tolerability, safety, and efficacy of 1200 mg NAC given to patients with MS over 15 months (Schoeps et al. 2022). The target primary outcomes include the degree of brain atrophy through MRI of brain lesions, quality of life, disability severity, limb function, and biomarkers such as neurofilament levels in the blood (Schoeps et al. 2022). A recent, randomized, placebo-controlled clinical trial explored the use of NAC in patients with MS at 600 mg of NAC twice a day for 8 weeks (Khalatbari Mohseni et al. 2023). The study measured serum levels of glutathione, NO, and malondialdehyde. Serum malondialdehyde was significantly decreased in the intervention group, but changes in glutathione and NO were not statistically significant between the treatment and control groups. The trial also measured anxiety levels through a Hospital Anxiety and Depression Scale and showed significant decreases in anxiety scores in the treatment compared to the control group (Khalatbari Mohseni et al. 2023). Given the positive results of NAC treatment in patients with MS, additional clinical trials of longer durations are needed to determine its efficacy and safety profile.

10.13 Peroxiredoxin 6 (PRDX6)

PRDX6 is an enzyme that protects cells from ROS and utilizes glutathione as a reducing agent to exert its antioxidative role (Fisher 2010). Its expression is predominantly in astrocytes, and it has been recently associated with improved spatial memory in a study on mice with or without the PRDX6 gene (Phasuk et al. 2021). Preclinical and clinical evidence of the efficacy of PRDX6 in MS is meager or even obsolete. A preclinical trial on mice with EAE used exogenous PRDX6 and thymulin, a peptide with immunomodulatory effects (Lunin et al. 2023). The mice were split into two groups: one group received 6 mg/kg of PRDX6 on days 2, 7, and 10 of the 22-day intervention, while the other group received a higher dose of 20 mg/kg. Thymulin was injected intraperitoneally at a 0.15 mg/kg concentration on alternate days. The study showed significant improvements in blood-brain barrier integrity and lymphocytic infiltration (Lunin et al. 2023). Although the combined PRDX6 and Thymulin treatment was deemed beneficial, additional research is needed to understand the effect of PRDX6 alone as well as its tolerability and safety profile.

10.14 Conclusion

In summary, exploring antioxidants as potential therapeutic interventions for MS stands as a promising frontier in the pursuit of effective treatment strategies. MS, characterized by inflammation, demyelination, and neurodegeneration, is intricately linked to oxidative stress, prompting a closer examination of compounds like Ginkgo biloba, melatonin, curcumin, edaravone, N-acetyl cysteine, Acer Tantrum Oil, and Ilex paraguariensis St. Hilaire, touted for their antioxidative properties.

These compounds exert their antioxidant effects through various mechanisms of action, encompassing anti-inflammatory, neuroprotective, and immune-modulating effects. Ginkgo biloba, in particular, reveals potential as an adjunctive therapy for MS. Acer Tantrum Oil, rich in polyunsaturated fatty acids, and Ilex paraguariensis St. Hilaire, known for its anti-inflammatory and antioxidant properties, emerge as novel candidates for further investigation.

While preclinical studies present promising avenues, the transition to more rigorous clinical trials is essential to establish the safety and efficacy of these antioxidants in managing MS. The complexities of MS demand a comprehensive approach, and the collective exploration of antioxidants underscores the need for nuanced insights into their potential benefits and risks. As we delve deeper into understanding the intricate relationship between oxidative stress and MS, these antioxidants offer hope for improved outcomes, potentially slowing disease progression and enhancing the quality of life for individuals with the disease. The road from bench research to bedside clinical application involves continued investigation, refinement of treatment protocols, and a thorough evaluation of long-term safety. As the collective efforts progress, antioxidants may find their place in the holistic care of those navigating the challenges posed by MS.

10.15 Summary

MS is a complex autoimmune disease characterized by inflammation, demyelination, and neurodegeneration, with oxidative stress playing a critical role in its progression. This chapter explores the therapeutic potential of specific antioxidants—Ginkgo biloba, melatonin, curcumin, Edaravone, and N-acetyl cysteine—in the context of MS. Each antioxidant offers unique benefits: Ginkgo biloba modulates immune responses and reduces neuroinflammation, melatonin displays neuroprotective and immunomodulatory effects, curcumin demonstrates anti-inflammatory and antioxidant actions, Edaravone alleviates oxidative stress, and N-acetyl cysteine enhances glutathione levels. Preclinical studies highlight their potential, but more rigorous clinical trials are needed to establish their efficacy and safety in MS management. The chapter underscores the need for continued research to optimize antioxidant therapies to slow disease progression and improve the quality of life for individuals with MS.

References

Abo Taleb HA, Alghamdi BS (2020) Neuroprotective effects of melatonin during demyelination and Remyelination stages in a mouse model of multiple sclerosis. J Mol Neurosci 70(3):386–402. https://doi.org/10.1007/s12031-019-01425-6

Adamczyk B, Adamczyk-Sowa M (2016) New insights into the role of oxidative stress mechanisms in the pathophysiology and treatment of multiple sclerosis. Oxidative Med Cell Longev 2016:1

Alkhatib A, Atcheson R (2017) Yerba Maté (Ilex paraguariensis) metabolic, satiety, and mood state effects at rest and during prolonged exercise. Nutrients 9(8):882

Atalay S, Jarocka-Karpowicz I, Skrzydlewska E (2019) Antioxidative and anti-inflammatory properties of Cannabidiol. Antioxidants (Basel) 9(1):21. https://doi.org/10.3390/antiox9010021

Atefi M, Darand M, Entezari MH, Jamialahmadi T, Bagheriniya M, Sahebkar A (2021) A systematic review of the clinical use of curcumin for the management of gastrointestinal diseases. In: Guest PC (ed) Studies on biomarkers and new targets in aging research in Iran: focus on turmeric and curcumin. Springer, Cham, pp 295–326. https://doi.org/10.1007/978-3-030-56153-6_18

Bakhtiari M, Ghasemi N, Salehi H, Amirpour N, Kazemi M, Mardani M (2021) Evaluation of Edaravone effects on the differentiation of human adipose derived stem cells into oligodendrocyte cells in multiple sclerosis disease in rats. Life Sci 282:119812. https://doi.org/10.1016/j.lfs.2021.119812

Bássoli RMF, Audi D, Ramalho BJ, Audi M, Quesada KR, Barbalho SM (2023) The effects of curcumin on neurodegenerative diseases: a systematic review. J Herb Med 42:100771. https://doi.org/10.1016/j.hermed.2023.100771

Bastos DHM, Beltrame D, Matsumoto RLT, Carvalho P, Ribeiro M (2007) Yerba maté: pharmacological properties, research and biotechnology. Med Arom Plant Sci Biotechnol 1:37–46

Bellmann-Strobl J, Paul F, Wuerfel J, Dörr J, Infante-Duarte C, Heidrich E, Körtgen B, Brandt A, Pfüller C, Radbruch H, Rust R, Siffrin V, Aktas O, Heesen C, Faiss J, Hoffmann F, Lorenz M, Zimmermann B, Groppa S, Zipp F (2021) Epigallocatechin Gallate in relapsing-remitting multiple sclerosis. Neurol Neuroimmunol Neuroinflam 8(3):e981. https://doi.org/10.1212/NXI.0000000000000981

Bittner F, Murchison C, Koop D, Bourdette D, Spain R (2017) Lipoic acid pharmacokinetics at baseline and 1 year in secondary progressive MS. Neurol Neuroimmunol Neuroinflamm 4(5):e380. https://doi.org/10.1212/nxi.0000000000000380

Brenneisen R, Egli A, Elsohly M, Henn V, Spiess Y (1996) The effect of orally and rectally administered Δ 9-tetrahydrocannabinol on spasticity: a pilot study with 2 patients. Int J Clin Pharmacol Ther 34:446–452

Cameron M, Taylor C, Lapidus J, Ramsey K, Koop D, Spain R (2020) Gastrointestinal tolerability and absorption of R-versus RS-lipoic acid in progressive multiple sclerosis: a randomized crossover trial. J Clin Pharmacol 60(8):1099–1106. https://doi.org/10.1002/jcph.1605

Camotti Bastos M, Cherobim VF, Reissmann CB, Fernandes Kaseker J, Gaiad S (2018) Yerba mate: nutrient levels and quality of the beverage depending on the harvest season. J Food Compos Anal 69:1–6. https://doi.org/10.1016/j.jfca.2018.01.019

Cha SJ, Kim K (2022) Effects of the Edaravone, a drug approved for the treatment of amyotrophic lateral sclerosis, on mitochondrial function and neuroprotection. Antioxidants 11(2):195

Cheminet G, Baroni MV, Wunderlin DA, Di Paola Naranjo RD (2021) Antioxidant properties and phenolic composition of "composed yerba mate". J Food Sci Technol 58(12):4711–4721. https://doi.org/10.1007/s13197-020-04961-x

Clifford DB (1983) Tetrahydrocannabinol for tremor in multiple sclerosis. Ann Neurol 13(6):669–671. https://doi.org/10.1002/ana.410130616

Collin C, Ehler E, Waberzinek G, Alsindi Z, Davies P, Powell K, Notcutt W, O'Leary C, Ratcliffe S, Nováková I, Zapletalova O, Piková J, Ambler Z (2010) A double-blind, randomized, placebo-controlled, parallel-group study of Sativex, in subjects with symptoms of spasticity due to multiple sclerosis. Neurol Res 32(5):451–459. https://doi.org/10.1179/016164109X12590518685660

Cuerda-Ballester M, Proaño B, Alarcón-Jimenez J, de Bernardo N, Villaron-Casales C, Lajara Romance JM, de la Rubia Ortí JE (2023) Improvements in gait and balance in patients with multiple sclerosis after treatment with coconut oil and epigallocatechin gallate. A pilot study [10.1039/D2FO02207A]. Food Funct 14(2):1062–1071. https://doi.org/10.1039/D2FO02207A

Dartora N, de Souza LM, Paiva SMM, Scoparo CT, Iacomini M, Gorin PAJ, Rattmann YD, Sassaki GL (2013) Rhamnogalacturonan from Ilex paraguariensis: a potential adjuvant in sepsis treatment. Carbohydr Polym 92(2):1776–1782. https://doi.org/10.1016/j.carbpol.2012.11.013

Dolati S, Aghebati-Maleki L, Ahmadi M, Marofi F, Babaloo Z, Ayramloo H, Jafarisavari Z, Oskouei H, Afkham A, Younesi V, Nouri M, Yousefi M (2018) Nanocurcumin restores aberrant miRNA expression profile in multiple sclerosis, randomized, double-blind, placebo-controlled trial. J Cell Physiol 233(7):5222–5230. https://doi.org/10.1002/jcp.26301

Dolati S, Babaloo Z, Ayromlou H, Ahmadi M, Rikhtegar R, Rostamzadeh D, Roshangar L, Nouri M, Mehdizadeh A, Younesi V (2019) Nanocurcumin improves regulatory T-cell frequency and function in patients with multiple sclerosis. J Neuroimmunol 327:15–21

Dong F, Yan W, Meng Q, Song X, Cheng B, Liu Y, Yao R (2023) Ebselen alleviates white matter lesions and improves cognitive deficits by attenuating oxidative stress via Keap1/Nrf2 pathway in chronic cerebral hypoperfusion mice. Behav Brain Res 448:114444. https://doi.org/10.1016/j.bbr.2023.114444

ELBini-Dhouib I, Manai M, Neili NE, Marzouki S, Sahraoui G, Ben Achour W, Zouaghi S, Ben Ahmed M, Doghri R, Srairi-Abid N (2022) Dual mechanism of action of curcumin in experimental models of multiple sclerosis. Int J Mol Sci 23(15):8658. https://doi.org/10.3390/ijms23158658

Farukhi ZM, Mora S, Manson JE (2021) Marine Omega-3 fatty acids and cardiovascular disease prevention: seeking clearer water. Mayo Clin Proc 96(2):277–279. https://doi.org/10.1016/j.mayocp.2020.12.013

Fernández-Ortiz M, Sayed RKA, Fernández-Martínez J, Cionfrini A, Aranda-Martínez P, Escames G, de Haro T, Acuña-Castroviejo D (2020) Melatonin/Nrf2/NLRP3 connection in mouse heart mitochondria during aging. Antioxidants 9(12):1187

Ferretti G, Bacchetti T, Principi F, Di Ludovico F, Viti B, Angeleri VA, Danni M, Provinciali L (2005) Increased levels of lipid hydroperoxides in plasma of patients with multiple sclerosis: a relationship with paraoxonase activity. Mult Scler J 11(6):677–682. https://doi.org/10.1191/1352458505ms1240oa

Fiedler SE, Yadav V, Kerns AR, Tsang C, Markwardt S, Kim E, Spain R, Bourdette D, Salinthone S (2018) Lipoic acid stimulates cAMP production in healthy control and secondary progressive MS subjects. Mol Neurobiol 55(7):6037–6049. https://doi.org/10.1007/s12035-017-0813-y

Filip R, Lotito S, Ferraro G, Fraga C (2000) Antioxidant activity of Ilex paraguariensis and related species. Nutr Res 20:1437–1446. https://doi.org/10.1016/S0271-5317(00)80024-X

Fisher AB (2010) Peroxiredoxin 6: a bifunctional enzyme with glutathione peroxidase and phospholipase A2 activities. Antioxid Redox Signal 15(3):831–844. https://doi.org/10.1089/ars.2010.3412

Herges K, Millward JM, Hentschel N, Infante-Duarte C, Aktas O, Zipp F (2011) Neuroprotective effect of combination therapy of Glatiramer acetate and Epigallocatechin-3-Gallate in Neuroinflammation. PLoS One 6(10):e25456. https://doi.org/10.1371/journal.pone.0025456

Heś M, Dziedzic K, Górecka D, Jędrusek-Golińska A, Gujska E (2019) Aloe vera (L.) Webb.: natural sources of antioxidants—a review. Plant Foods Hum Nutr 74(3):255–265. https://doi.org/10.1007/s11130-019-00747-5

Johnson SK, Diamond BJ, Rausch S, Kaufman M, Shiflett SC, Graves L (2006) The effect of ginkgo biloba on functional measures in multiple sclerosis: a pilot randomized controlled trial. Explore 2(1):19–24. https://doi.org/10.1016/j.explore.2005.10.007

Kawakita E, Hashimoto M, Shido O (2006) Docosahexaenoic acid promotes neurogenesis in vitro and in vivo. Neuroscience 139(3):991–997. https://doi.org/10.1016/j.neuroscience.2006.01.021

Khalatbari Mohseni G, Hosseini SA, Majdinasab N, Cheraghian B (2023) Effects of N-acetylcysteine on oxidative stress biomarkers, depression, and anxiety symptoms in patients with multiple sclerosis. Neuropsychopharmacol Rep 43(3):382–390. https://doi.org/10.1002/npr2.12360

Khalili M, Eghtesadi S, Mirshafiey A, Eskandari G, Sanoobar M, Sahraian MA, Motevalian A, Norouzi A, Moftakhar S, Azimi A (2014) Effect of lipoic acid consumption on oxidative stress among multiple sclerosis patients: a randomized controlled clinical trial. Nutr Neurosci 17(1):16–20. https://doi.org/10.1179/1476830513Y.0000000060

Khalili M, Soltani M, Moghadam SA, Dehghan P, Azimi A, Abbaszadeh O (2017) Effect of alpha-lipoic acid on asymmetric dimethylarginine and disability in multiple sclerosis patients: a randomized clinical trial. Electron Physician 9(7):4899–4905. https://doi.org/10.19082/4899

Kim YH, Oh TW, Park E, Yim NH, Park KI, Cho WK, Ma JY (2018) Anti-inflammatory and anti-apoptotic effects of acer Palmatum thumb. Extract, KIOM-2015EW, in a hyperosmolar-stress-induced in vitro dry eye model. Nutrients 10(3):282. https://doi.org/10.3390/nu10030282

Klingenberg R, Gerdes N, Badeau RM, Gisterå A, Strodthoff D, Ketelhuth DFJ, Lundberg AM, Rudling M, Nilsson SK, Olivecrona G, Zoller S, Lohmann C, Lüscher TF, Jauhiainen M,

Sparwasser T, Hansson GK (2013) Depletion of FOXP3+ regulatory T cells promotes hypercholesterolemia and atherosclerosis. J Clin Invest 123(3):1323–1334. https://doi.org/10.1172/jci63891

Kristina BS, Troels SJ, Flemming WB (2004) Does the cannabinoid dronabinol reduce central pain in multiple sclerosis? Randomised double blind placebo controlled crossover trial. BMJ 329(7460):253. https://doi.org/10.1136/bmj.38149.566979.AE

Kungel P, Correa VG, Corrêa RCG, Peralta RA, Soković M, Calhelha RC, Bracht A, Ferreira I, Peralta RM (2018) Antioxidant and antimicrobial activities of a purified polysaccharide from yerba mate (Ilex paraguariensis). Int J Biol Macromol 114:1161–1167. https://doi.org/10.1016/j.ijbiomac.2018.04.020

Licht-Mayer S, Wimmer I, Traffehn S, Metz I, Brück W, Bauer J, Bradl M, Lassmann H (2015) Cell type-specific Nrf2 expression in multiple sclerosis lesions. Acta Neuropathol 130(2):263–277. https://doi.org/10.1007/s00401-015-1452-x

Lourenço SC, Moldão-Martins M, Alves VD (2019) Antioxidants of natural plant origins: from sources to food industry applications. Molecules 24(22):4132. https://doi.org/10.3390/molecules24224132

Lovera JF, Kim E, Heriza E, Fitzpatrick M, Hunziker J, Turner AP, Adams J, Stover T, Sangeorzan A, Sloan A, Howieson D, Wild K, Haselkorn J, Bourdette D (2012) Ginkgo biloba does not improve cognitive function in MS. Neurology 79(12):1278–1284. https://doi.org/10.1212/WNL.0b013e31826aac60

Loy BD, Fling BW, Horak FB, Bourdette DN, Spain RI (2018) Effects of lipoic acid on walking performance, gait, and balance in secondary progressive multiple sclerosis. Complement Ther Med 41:169–174. https://doi.org/10.1016/j.ctim.2018.09.006

Lu F, Selak M, O'Connor J, Croul S, Lorenzana C, Butunoi C, Kalman B (2000) Oxidative damage to mitochondrial DNA and activity of mitochondrial enzymes in chronic active lesions of multiple sclerosis. J Neurol Sci 177(2):95–103. https://doi.org/10.1016/s0022-510x(00)00343-9

Lunin SM, Novoselova EG, Glushkova OV, Parfenyuk SB, Kuzekova AA, Novoselova TV, Sharapov MG, Mubarakshina EK, Goncharov RG, Khrenov MO (2023) Protective effect of exogenous peroxiredoxin 6 and thymic peptide thymulin on BBB conditions in an experimental model of multiple sclerosis. Arch Biochem Biophys 746:109729. https://doi.org/10.1016/j.abb.2023.109729

Mähler A, Mandel S, Lorenz M, Ruegg U, Wanker EE, Boschmann M, Paul F (2013) Epigallocatechin-3-gallate: a useful, effective and safe clinical approach for targeted prevention and individualised treatment of neurological diseases? EPMA J 4(1):5. https://doi.org/10.1186/1878-5085-4-5

Mähler A, Steiniger J, Bock M, Klug L, Parreidt N, Lorenz M, Zimmermann BF, Krannich A, Paul F, Boschmann M (2015) Metabolic response to epigallocatechin-3-gallate in relapsing-remitting multiple sclerosis: a randomized clinical trial 234. Am J Clin Nutr 101(3):487–495. https://doi.org/10.3945/ajcn.113.075309

Markovà J, Essner U, Akmaz B, Marinelli M, Trompke C, Lentschat A, Vila C (2019) Sativex® as add-on therapy vs. further optimized first-line ANTispastics (SAVANT) in resistant multiple sclerosis spasticity: a double-blind, placebo-controlled randomised clinical trial. Int J Neurosci 129(2):119–128. https://doi.org/10.1080/00207454.2018.1481066

Martins APL, & Reissmann CB (2007) Material vegetal e as rotinas laboratoriais nos procedimentos químico-analíticos. Scientia Agraria 8(1):1–17

Masoomeh Y, Gholamhassan V, Ali Akbar M, Fardin F, Vida H (2019) A randomized control trial study to determine the effect of melatonin on serum levels of IL-1β and TNF-α in patients with multiple sclerosis. Iran J Allergy Asthma Immunol 18(6):649. https://doi.org/10.18502/ijaai.v18i6.2177

Meinck HM, Schönle PW, Conrad B (1989) Effect of cannabinoids on spasticity and ataxia in multiple sclerosis. J Neurol 236(2):120–122. https://doi.org/10.1007/BF00314410

Miller ED, Dziedzic A, Saluk-Bijak J, Bijak M (2019) A review of various antioxidant compounds and their potential utility as complementary therapy in multiple sclerosis. Nutrients 11(7):1528

Mojaverrostami S, Bojnordi MN, Ghasemi-Kasman M, Ebrahimzadeh MA, Hamidabadi HG (2018) A review of herbal therapy in multiple sclerosis. Adv Pharm Bull 8(4):575–590. https://doi.org/10.15171/apb.2018.066

Monti DA, Zabrecky G, Leist TP, Wintering N, Bazzan AJ, Zhan T, Newberg AB (2020) N-acetyl cysteine administration is associated with increased cerebral glucose metabolism in patients with multiple sclerosis: an exploratory study. Front Neurol 11:88. https://doi.org/10.3389/fneur.2020.00088

Moriya M, Nakatsuji Y, Miyamoto K, Okuno T, Kinoshita M, Kumanogoh A, Kusunoki S, Sakoda S (2008) Edaravone, a free radical scavenger, ameliorates experimental autoimmune encephalomyelitis. Neurosci Lett 440(3):323–326. https://doi.org/10.1016/j.neulet.2008.05.110

Muñoz-Jurado A, Escribano BM, Caballero-Villarraso J, Galván A, Agüera E, Santamaría A, Túnez I (2022) Melatonin and multiple sclerosis: antioxidant, anti-inflammatory and immunomodulator mechanism of action. Inflammopharmacology 30(5):1569–1596. https://doi.org/10.1007/s10787-022-01011-0

Nabizadeh F (2023) Antioxidant supplements and cognition in multiple sclerosis: a systematic review. Neurology 100:P14–3.003. https://doi.org/10.1212/WNL.0000000000202806

Nakamura Y, Feng Q, Kumagai T, Torikai K, Ohigashi H, Osawa T, Noguchi N, Niki E, Uchida K (2002) Ebselen, a glutathione peroxidase mimetic Seleno-organic compound, as a multifunctional antioxidant: implication for inflammation-associated carcinogenesis*. J Biol Chem 277(4):2687–2694. https://doi.org/10.1074/jbc.M109641200

Nouh RA, Kamal A, Abdelnaser A (2023) Cannabinoids and multiple sclerosis: a critical analysis of therapeutic potentials and safety concerns. Pharmaceutics 15(4):1151

Novotna A, Mares J, Ratcliffe S, Novakova I, Vachova M, Zapletalova O, Gasperini C, Pozzilli C, Cefaro L, Comi G, Rossi P, Ambler Z, Stelmasiak Z, Erdmann A, Montalban X, Klimek A, Davies P, Group, t. S. S. S (2011) A randomized, double-blind, placebo-controlled, parallel-group, enriched-design study of nabiximols* (Sativex®), as add-on therapy, in subjects with refractory spasticity caused by multiple sclerosis. Eur J Neurol 18(9):1122–1131. https://doi.org/10.1111/j.1468-1331.2010.03328.x

Parlapani E, Schmitt A, Erdmann A, Bernstein H-G, Breunig B, Gruber O, Petroianu G, von Wilmsdorff M, Schneider-Axmann T, Honer W, Falkai P (2009) Association between myelin basic protein expression and left entorhinal cortex pre-alpha cell layer disorganization in schizophrenia. Brain Res 1301:126–134. https://doi.org/10.1016/j.brainres.2009.09.007

Patti F, Messina S, Solaro C, Amato MP, Bergamaschi R, Bonavita S, Bossio RB, Morra VB, Costantino GF, Cavalla P, Centonze D, Comi G, Cottone S, Danni M, Francia A, Gajofatto A, Gasperini C, Ghezzi A, Iudice A, Zappia M (2016) Efficacy and safety of cannabinoid oromucosal spray for multiple sclerosis spasticity. J Neurol Neurosurg Psychiatr 87(9):944. https://doi.org/10.1136/jnnp-2015-312591

Petracca M, Quarantelli M, Moccia M, Vacca G, Satelliti B, D'Ambrosio G, Carotenuto A, Ragucci M, Assogna F, Capacchione A, Lanzillo R, Morra VB (2021) ProspeCtive study to evaluate efficacy, safety and tolerability of dietary suppleMeNT of curcumin (BCM95) in subjects with active relapsing MultIple sclerosis treated with subcutaNeous interferon beta 1a 44 mcg TIW (CONTAIN): a randomized, controlled trial. Mult Scler Relat Disord 56:103274. https://doi.org/10.1016/j.msard.2021.103274

Petro DJ, Ellenberger C Jr (1981) Treatment of human spasticity with Δ9-tetrahydrocannabinol. J Clin Pharmacol 21(S1):413S–416S. https://doi.org/10.1002/j.1552-4604.1981.tb02621.x

Phasuk S, Jasmin S, Pairojana T, Chang H-K, Liang K-C, Liu IY (2021) Lack of the peroxiredoxin 6 gene causes impaired spatial memory and abnormal synaptic plasticity. Mol Brain 14(1):72. https://doi.org/10.1186/s13041-021-00779-6

Plemel JR, Juzwik CA, Benson CA, Monks M, Harris C, Ploughman M (2015) Over-the-counter antioxidant therapies for use in multiple sclerosis: a systematic review. Mult Scler J 21(12):1485–1495. https://doi.org/10.1177/1352458515601513

Pulido-Moran M, Moreno-Fernandez J, Ramirez-Tortosa C, Ramirez-Tortosa M (2016) Curcumin and health. Molecules 21(3):264. https://www.mdpi.com/1420-3049/21/3/264

Qi X, Lewin AS, Sun L, Hauswirth WW, Guy J (2006) Mitochondrial protein nitration primes neurodegeneration in experimental autoimmune encephalomyelitis. J Biol Chem 281(42):31950–31962. https://doi.org/10.1074/jbc.M603717200

Roelcke U, Kappos L, Lechner-Scott J, Brunnschweiler H, Huber S, Ammann W, Plohmann A, Dellas S, Maguire RP, Missimer J, Radii EW, Steck A, Leenders KL (1997) Reduced glucose metabolism in the frontal cortex and basal ganglia of multiple sclerosis patients with fatigue. Neurology 48(6):1566–1571. https://doi.org/10.1212/WNL.48.6.1566

Rog DJ, Nurmikko TJ, Friede T, Young CA (2005) Randomized, controlled trial of cannabis-based medicine in central pain in multiple sclerosis. Neurology 65(6):812–819. https://doi.org/10.1212/01.wnl.0000176753.45410.8b

Rust R, Chien C, Scheel M, Brandt AU, Dörr J, Wuerfel J, Klumbies K, Zimmermann H, Lorenz M, Wernecke K-D, Bellmann-Strobl J, Paul F (2021) Epigallocatechin Gallate in progressive MS. Neurol Neuroimmunol Neuroinflam 8(3):e964. https://doi.org/10.1212/NXI.0000000000000964

Salinthone S, Yadav V, Schillace RV, Bourdette DN, Carr DW (2010) Lipoic acid attenuates inflammation via cAMP and protein kinase a signaling. PLoS One 5(9):e13058. https://doi.org/10.1371/journal.pone.0013058

Salinthone S, Yadav V, Ganesh M, Cherala G, Shinto L, Koop D, Bourdette D, Carr D (2012) P 02.183. Comparing the bioavailability of two forms of lipoic acid in multiple sclerosis. BMC Complement Altern Med 12(1):P239. https://doi.org/10.1186/1472-6882-12-S1-P239

Schoeps VA, Graves JS, Stern WA, Zhang L, Nourbakhsh B, Mowry EM, Henry RG, Waubant E (2022) N-acetyl cysteine as a neuroprotective agent in progressive multiple sclerosis (NACPMS) trial: study protocol for a randomized, double-blind, placebo-controlled add-on phase 2 trial. Contemp Clin Trials 122:106941. https://doi.org/10.1016/j.cct.2022.106941

Singh SK, Srivastav S, Castellani RJ, Plascencia-Villa G, Perry G (2019) Neuroprotective and antioxidant effect of Ginkgo biloba extract against AD and other neurological disorders. Neurotherapeutics 16(3):666–674. https://doi.org/10.1007/s13311-019-00767-8

Siotto M, Filippi MM, Simonelli I, Landi D, Ghazaryan A, Vollaro S, Ventriglia M, Pasqualetti P, Rongioletti MCA, Squitti R, Vernieri F (2019) Oxidative stress related to iron metabolism in relapsing remitting multiple sclerosis patients with low disability [original research]. Front Neurosci 13:86. https://doi.org/10.3389/fnins.2019.00086

Sochocka M, Ochnik M, Sobczyński M, Gębura K, Zambrowicz A, Naporowski P, Leszek J (2022) Ginkgo biloba leaf extract improves an innate immune response of peripheral blood leukocytes of Alzheimer's disease patients. Nutrients 14(10):2022. https://doi.org/10.3390/nu14102022

Sorosina M, Clarelli F, Ferrè L, Osiceanu AM, Unal NT, Mascia E, Martinelli V, Comi G, Benigni F, Esposito F, Martinelli Boneschi F (2018) Clinical response to Nabiximols correlates with the downregulation of immune pathways in multiple sclerosis. Eur J Neurol 25(7):934–e970. https://doi.org/10.1111/ene.13623

Spain R, Powers K, Murchison C, Heriza E, Winges K, Yadav V, Cameron M, Kim E, Horak F, Simon J, Bourdette D (2017) Lipoic acid in secondary progressive MS. Neurol Neuroimmunol Neuroinflam 4(5):e374. https://doi.org/10.1212/NXI.0000000000000374

Tobore TO (2021) Oxidative/Nitroxidative stress and multiple sclerosis. J Mol Neurosci 71(3):506–514. https://doi.org/10.1007/s12031-020-01672-y

Ungerleider JT, Andyrsiak T, Fairbanks L, Ellison GW, Myers LW (1988) Delta-9-THC in the treatment of spasticity associated with multiple sclerosis. Adv Alcohol Subst Abuse 7(1):39–50. https://doi.org/10.1300/J251v07n01_04

Villar-Delfino PH, Gomes NAO, Christo PP, Nogueira-Machado JA, Volpe CMO (2022) Edaravone inhibits the production of reactive oxygen species in phagocytosis-and PKC-stimulated granulocytes from multiple sclerosis patients Edaravone modulate oxidative stress in multiple sclerosis. J Central Nerv Syst Dis 14:11795735221092524. https://doi.org/10.1177/11795735221092524

Wade DT, Robson P, House H, Makela P, Aram J (2003) A preliminary controlled study to determine whether whole-plant cannabis extracts can improve intractable neurogenic symptoms. Clin Rehabil 17(1):21–29. https://doi.org/10.1191/0269215503cr581oa

Xie H, Yang X, Cao Y, Long X, Shang H, Jia Z (2022) Role of lipoic acid in multiple sclerosis. CNS Neurosci Ther 28(3):319–331. https://doi.org/10.1111/cns.13793

Xue Y, Zhu X, Yan W, Zhang Z, Cui E, Wu Y, Li C, Pan J, Yan Q, Chai X (2022) Dietary supplementation with Acer truncatum oil promotes Remyelination in a mouse model of multiple sclerosis. Front Neurosci 16:860280

Yadav V, Marracci G, Lovera J, Woodward W, Bogardus K, Marquardt W, Shinto L, Morris C, Bourdette D (2005) Lipoic acid in multiple sclerosis: a pilot study. Mult Scler J 11(2):159–165. https://doi.org/10.1191/1352458505ms1143oa

Yadav V, Marracci GH, Munar MY, Cherala G, Stuber LE, Alvarez L, Shinto L, Koop DR, Bourdette DN (2010) Pharmacokinetic study of lipoic acid in multiple sclerosis: comparing mice and human pharmacokinetic parameters. Mult Scler J 16(4):387–397. https://doi.org/10.1177/1352458509359722

Yosefi-Fard M, Vaezi G, Maleki-Rad AA, Faraji F, Hojati V (2020) Effect of melatonin on serum levels of INF-1β and vitamin B12 in patients with multiple sclerosis: a randomized controlled trial. IJT 14(1):19–24. https://doi.org/10.32598/ijt.14.1.19

Zajicek J, Fox P, Sanders H, Wright D, Vickery J, Nunn A, Thompson A (2003) Cannabinoids for treatment of spasticity and other symptoms related to multiple sclerosis (CAMS study): multicentre randomised placebo-controlled trial. Lancet 362(9395):1517–1526. https://doi.org/10.1016/S0140-6736(03)14738-1

Zajicek JP, Sanders HP, Wright DE, Vickery PJ, Ingram WM, Reilly SM, Nunn AJ, Teare LJ, Fox PJ, Thompson AJ (2005) Cannabinoids in multiple sclerosis (CAMS) study: safety and efficacy data for 12 months follow up. J Neurol Neurosurg Psychiatr 76(12):1664. https://doi.org/10.1136/jnnp.2005.070136

Zhang Y, Bhavnani BR (2006) Glutamate-induced apoptosis in neuronal cells is mediated via caspase-dependent and independent mechanisms involving calpain and caspase-3 proteases as well as apoptosis inducing factor (AIF) and this process is inhibited by equine estrogens. BMC Neurosci 7(1):49. https://doi.org/10.1186/1471-2202-7-49

Zoi V, Galani V, Lianos GD, Voulgaris S, Kyritsis AP, Alexiou GA (2021) The role of curcumin in cancer treatment. Biomedicines 9(9):1086. https://doi.org/10.3390/biomedicines9091086

Chapter 11
Dietary Regimens: Whole Grains and Multiple Sclerosis

Haia M. R. Abdulsamad, Amna Baig, Sara Aljoudi, Nadia Rabeh, Zakia Dimassi, and Hamdan Hamdan

Abstract For decades, disease-modifying therapy has been the central approach in managing multiple sclerosis (MS), focusing on restoring motor function, preventing disease relapse, and maintaining remission. However, MS patients continue to require ways to optimize their lifestyle and diet to cope with the accompanying fatigue and psychological comorbidities. This chapter investigates the multifaceted impact of dietary interventions, particularly emphasizing the role of whole grains and fibers in managing MS. Exploring evidence from experimental models, the chapter explores the potential of fibers like guar gum and cellulose in ameliorating MS symptoms. Emphasis is placed on the modulation of gut microbiota and inflammatory markers, offering insights into the broader health benefits of whole grains, including cognitive function improvement and reduced risks of various diseases.

Keywords Multiple sclerosis (MS) · Whole grains · Fibers · Neuromyelitis optica (NMO) · Experimental autoimmune encephalomyelitis (EAE)

Abbreviations

BBB	Blood-brain barrier
Ccr10	C-C chemokine receptor type 10
Ccr2	C-C chemokine receptor type 2

H. M. R. Abdulsamad · A. Baig · S. Aljoudi · N. Rabeh · H. Hamdan (✉)
Department of Biological Sciences, College of Medicine and Health Sciences, Khalifa University, Abu Dhabi, United Arab Emirates
e-mail: hamdan.hamdan@ku.ac.ae

Z. Dimassi
Department of Medical Sciences, College of Medicine and Health Sciences, Khalifa University, Abu Dhabi, United Arab Emirates

© The Author(s), under exclusive license to Springer Nature Singapore Pte Ltd. 2024
H. Hamdan (ed.), *Exploring the Effects of Diet on the Development and Prognosis of Multiple Sclerosis (MS)*, Nutritional Neurosciences, https://doi.org/10.1007/978-981-97-4673-6_11

Ccr9	C-C chemokine receptor type 9
CNS	Central nervous system
CRP	C-reactive protein
CSF	Cerebrospinal fluid
Cxcr1	C-X-C motif chemokine receptor 1
Cxcr3	C-X-C motif chemokine receptor 3
EAE	Experimental autoimmune encephalomyelitis
ICAM-1	Intercellular adhesion molecule–1
IFN	Interferon
IL	Interleukin
LFA-1	Leukocyte function-associated antigen-1
MD	Mediterranean diet
MS	Multiple sclerosis
NF-κB	Nuclear factor kappa B
NMO	Neuromyelitis optica
OSE	Optico-spinal encephalomyelitis
PAH	Pulmonary arterial hypertension
PBMCs	Peripheral blood mononuclear cells
PDDS	Patient-determined disease steps
PHGG	Partially hydrolyzed guar gum
Th	T helper
TNF-α	Tumor necrosis factor alpha
VCAM-1	Vascular cell adhesion molecule-1
VLA-4	Very late antigen 4

Learning Objectives
- Describe the significance of dietary interventions, particularly whole grains and fibers, in managing MS symptoms and progression.
- Identify the specific fibers, such as guar gum and cellulose, and their potential to alleviate MS symptoms through experimental model findings.
- Explain how dietary fibers modulate gut microbiota and inflammatory markers, contributing to broader health benefits for MS patients.
- Evaluate the evidence linking whole grain consumption to improved cognitive function and reduced risks of neurological and other diseases in MS patients.
- Discuss the overall health benefits of whole grains, including their roles in reducing inflammation, enhancing insulin sensitivity, and lowering the risk of chronic diseases.

11.1 Introduction to the Role of Whole Grain Diets in Multiple Sclerosis

Whole grains are rich in both soluble and insoluble fiber. They include cereals, such as wheat, rye, oats, barley, and brown rice (Nirmala Prasadi and Joye 2020). The positive impact of dietary fiber on individuals with MS is well-documented in the literature. Studies reveal that fiber intake is associated with a reduction in relapse episodes and disability scores, as well as a decrease in circulating pro-inflammatory interleukin (IL)-17-producing T CD4+ lymphocytes when compared to individuals adhering to a Westernized diet (Saresella et al. 2017), findings that are indicative of an overall improvement in well-being and quality of life (Evers et al. 2022; Fitzgerald et al. 2018; Saresella et al. 2017). In subsequent chapters, we discuss how the Mediterranean diet (MD) contributes to alleviating the progression of MS, particularly through its emphasis on limited consumption of saturated fats, which are known to activate pro-inflammatory toll-like receptors.

Jahromi et al. (2012) found that a traditional diet rich in whole grains, legumes, low-fat dairy products, red meat, and vegetables was inversely proportional with the risk of relapsing/remitting MS (odds ratio = 0.15; 95% confidence interval: 0.03–0.18; $P = 0.028$). In contrast, diets low in whole grains and high in animal fats were linked to an increased risk of MS. These findings suggest that specific dietary patterns may influence the risk of MS (Jahromi et al. 2012).

In a study involving five groups of experimental autoimmune encephalomyelitis (EAE), a T cell-driven mouse model of MS, each group was supplemented with diets differing only in the content of fermentable fibers (30% guar gum, 30% pectin, 30% resistant starch, 30% inulin, or 0% fibers) for 14 days pre- and post-EAE exposure, while the control group was given a 5% cellulose diet (Fig. 11.1) (Fettig et al. 2022). Guar gum, a fiber isolated from guar beans (*Cyamopsis tetragonoloba*), comprises repeated units of galactose side chains with a mannopyranose backbone (Mudgil et al. 2014), effectively resisted CD4 T cell migratory potential, reduced CD4+ T helper (Th)-1 cells activation, delayed the onset of neuroinflammation, and

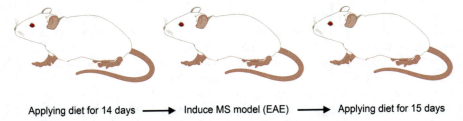

Applying diet for 14 days ⟶ Induce MS model (EAE) ⟶ Applying diet for 15 days

Fig. 11.1 A schematic diagram representing the experimental setup of Fettig et al. Five groups of EAE-mouse models were used; each was supplemented with diets that differed only in the content of fermentable fibres for 14 days pre- and 15 days post-EAE exposure. The diets were one of the following groups; 30% guar gum, 30% pectin, 30% resistant starch, 30% inulin and 0% fibres. The control group was fed a 5% cellulose diet (Fettig et al. 2022). *EAE* experimental autoimmune encephalomyelitis

mitigated symptom severity and MS incidence after EAE induction—actions specifically associated with Th1 activation and polarization. However, guar gum did not affect the Th17 cell control over interferon (IFN)-γ production. Histological findings also showed decreased inflammatory cell infiltration and demyelination in the spinal cord (Fettig et al. 2022).

Guar gum impeded CD4 T cells from penetrating the blood-brain barrier (BBB) by interfering with the expression of multiple integrins and chemokine receptors. Specifically, C-C chemokine receptor type 2 (Ccr2), C-C chemokine receptor type 10 (Ccr10), and C-X-C motif chemokine receptor 3 (Cxcr3) were downregulated. In contrast, C-C chemokine receptor type 9 (Ccr9) and C-X-C motif chemokine receptor 1 (Cxcr1), which are gut-homing-associated chemokine receptor genes, were upregulated (Fettig et al. 2022). Interestingly, MS patients show reduced CCR9 in peripheral blood mononuclear cells (PBMCs) and increased CXCR3+ CD4 T cells in cerebrospinal fluid (CSF) and central nervous system (CNS) lesions, indicating that guar gum can counteract the inflammatory changes seen in MS. Furthermore, integrins expressed on leukocyte membranes play a crucial role in their ability to penetrate the BBB. Guar gum mice showed decreased expression of the following CDs and their corresponding genes in leukocytes, including CD29 *(Itgb1)*, CD49b *(Itga2)*, CD49d *(Itga4)*, CD61 *(Itgb3)*, CD103 *(Itgae)*, and the *Itgb7 gene*. The reduced expression of the noted CDs affects the interaction of leukocyte function-associated antigen-1 (LFA-1) with intercellular adhesion molecule–1 (ICAM-1) and very late antigen 4 (VLA-4) with vascular cell adhesion molecule-1 (VCAM-1), as LFA-1 and VLA-4 are made of integrins αL/CD11a, β2, α4/CD49d, and β1, respectively. As a result, the migration of leukocytes to inflamed tissue in the brain is compromised (Fettig et al. 2022; Kadowaki et al. 2019; Sørensen et al. 1999, 2002).

11.2 Microbiota Modulation by Dietary Fibers and Immunomodulatory Potential

The reported influence of fibers on health may be attributed to the end product of their metabolism, namely short-chain fatty acids (Hung and Suzuki 2016), which possess the ability to modulate inflammatory environments as seen in some gastrointestinal autoimmune diseases, including inflammatory bowel disease and colitis (Horii et al. 2016; Hung and Suzuki 2016; Naito et al. 2006; Takagi et al. 2016). Short-chain fatty acids also play a role in the development and progression of cardiovascular diseases. For example, partially hydrolyzed guar gum (PHGG) treatment was found to suppress the development of pulmonary arterial hypertension (PAH) and vascular remodeling. This effect is attributed to PHGG's ability to enhance the production of short-chain fatty acids by gut microbiota, thereby protecting intestinal epithelial cells from damage. Gut dysbiosis has been observed in PAH animal models and PAH patients, suggesting a compromised gut barrier function that allows entry into the systemic circulation of toxic agents (Callejo et al. 2018; Kim et al. 2020) such as endotoxin and trimethylamine N-oxide, which may induce systemic inflammation and contribute to the development of pulmonary vascular remodeling in PAH. We conclude that PHGG, with its impact on short-chain fatty acids, emerges as a potential prebiotic agent for mitigating pulmonary arterial remodeling and hypertension (Sanada et al. 2023).

Previous research has shown that the transplantation of the microbiome from MS mice into healthy mice induced disease development in the recipients, indicating a possible role of gut dysbiosis in MS pathogenesis (Ochoa-Repáraz et al. 2018). Berer et al. conducted a study to explore the impact of dietary fibers on the microbiota, following an experimental design, genetically engineered mice were used to induce spontaneous EAE and optico-spinal encephalomyelitis (OSE) and were randomized to either a standard diet or a fiber-rich diet containing 26% cellulose. Mice fed the cellulose-rich diet exhibited a 23% incidence of spontaneous EAE, accompanied by a delayed onset of neurological symptoms. Conversely, the incidence of spontaneous EAE in mice on a standard diet was 55%, with no significant difference in disease severity between the two groups (Berer et al. 2018). To investigate the mechanism behind these findings, CD4+ T cells were extracted from the intestinal lamina propria to examine the cytokine profile in these mice. The observed effect in the cellulose group, leading to a decrease in disease incidence and a delay in symptom onset, was found to occur through the regulation of Th1 and Th2 cell activity. At the molecular level, the intervention group exhibited a reduced T cell response, evident by decreased expression of the transcription factor T-bet and the Th1 cell-cytokine IFN-γ in CD4+ T cells of the intestine. In the cellulose-rich group, there was a 50% reduction in IFN-γ-producing Th1 cells, while Th2 cells showed increased release of IL-4 and IL-5. The same trend in Th1 expression was observed in the spleen (Berer et al. 2018). In a similar fashion to guar gum, cellulose exerted no effect on Th17 cells (Berer et al. 2018; Fettig et al. 2022). Investigating the recruitment of innate immunity cells by Th1 and Th2 cells, the cellulose-rich group displayed a reduced neutrophil population in the spleen and lamina propria, coupled with an increase in eosinophils compared to the control group. This shift in cell population attenuated the effect of Th1 and heightened the impact of Th2, ultimately resulting in an altered immune response (Berer et al. 2018).

The effect on the microbiota was investigated by analyzing the fecal gut microbiome of the two groups, revealing growth in the families of Ruminococcaceae, Helicobacteraceae and Enterococcaceae, alongside a reduction in the families of Sutterellaceae, Lactobacillaceae and Coriobacteriaceae in the cellulose-rich group. At the genus level, there was an elevation in *Helicobacter, Enterococcus, Desulfovibrio, Parabacteroides, Pseudoflavonifractor, and Oscillibacter*, while *Lactobacillus, Parasutterella, Coprobacillus, and TM7 genera Incertae Sedis* were significantly reduced in the OSE model (Berer et al. 2018). Compared to EAE mice models, MS patients exhibited an abundance of pro-inflammatory bacteria, such as *Collinsella* and *Eggerthella*, alongside a decreased presence of beneficial bacteria, such as *Prevotella* and *Akkermansia* (Kurowska et al. 2023).

A comprehensive explanation of the observed changes in microbiota composition and the complex mechanisms underlying microbiota dysbiosis in MS patients is detailed in Chap. 10.

11.3 Whole Grains and Neuromyelitis Optica

Neuromyelitis Optica (NMO) is an autoantibody-mediated demyelinating disease that has a clinical presentation similar to MS (Lalan et al. 2012). Both conditions exhibit elevated Nuclear factor kappa B (NF-κB) and other inflammatory mediators, potentially modulated by whole grains (Rezaeimanesh et al. 2021). Whole grains and legumes regulate inflammatory cells and markers in NMO, influencing factors such as Th17 cells, IL-6, IL-17, tumor necrosis factor alpha (TNF-α), and IFN-γ, all of which play a role in MS (Rezaeimanesh et al. 2021). Increased consumption of whole grains and legumes by NMO patients showed a dose-dependent reduction in disease prevalence. This reduction was attributed to the total antioxidant capacity of whole grains and the low carbohydrate diet score associated with legumes. Legumes, rich in soluble fiber, complex carbohydrates, and polyphenols, exert antioxidant activity that could modulate the inflammatory cascade (Rezaeimanesh et al. 2021). Legumes' content of soluble (pectin, mucilage, gum) and insoluble (cellulose, hemicellulose) fibers, resistant starch, and oligosaccharides leads to increased populations of Lactobacillus and Bifidobacterium. Similarly, whole grains containing indigestible fiber undergo fermentation by gut microbiota, producing short-chain fatty acids and altering the microbiome by increasing Clostridium leptum (Rezaeimanesh et al. 2021). Therefore, we can conclude that whole grains' positive effects on MS are mediated through the modulation of microbiota and the inflammatory environment. The size or thickness of grains seems to play a role in influencing gut microbiota, such that thicker flakes significantly increase bifidobacteria compared to thin flakes, arguably due to the available surface area for digestive enzymes to act on resistant starch. Thicker flakes provide less surface area for enzymes, protecting resistant starch and allowing it to exert its effects on the microbiota, increasing the population of Bifidobacteria. This study suggests that individuals with MS incorporating whole grains into their diet should choose thicker flakes (Connolly et al. 2010).

A cross-sectional study involving 6989 patients with diagnosed MS looked into the patients' diet quality and determined that higher-quality diet components were associated with lower disability, as measured on the Patient-Determined Disease Steps (PDDS) and depression scores (Fitzgerald et al. 2018). The study defined a high-quality diet as comprising fruits, vegetables, legumes, and whole grains, with low intakes of sugar and red meat. Notably, among the high-quality diets, those composed of a high percentage of whole grains showed a lower rate of severe disease and disability.

11.4 Whole Grain and Cognitive Function

Chan et al. (2013) demonstrated that increased consumption of whole grains, combined with low-fat dairy, resulted in a lower rate of cognitive impairment, specifically Alzheimer's disease, in the Chinese population (Chan et al. 2013). Among the

Asian population, consumption of diet rich in whole grains was associated with higher scores on logical memory and global cognitive assessment scores, attributed to the elevated polyphenol content in whole grains (Goufo and Trindade 2017; Rajaram et al. 2019).

11.5 Whole Grain and Inflammation

Beyond the nervous system and neurocognitive function, the consumption of whole grains has been proven to decrease the risk of colon and rectum cancer, type 2 diabetes, and cardiovascular diseases (Montonen et al. 2013). It improves insulin sensitivity by inhibiting inflammation and oxidative stress, both important factors in the pathogenesis of MS. An interesting association was found, revealing that individuals who consume high-quality whole grains tend to be non-smokers, engage in sports and biking, and have higher levels of education (Montonen et al. 2013). Whole grains also reduce inflammatory markers such as C-reactive protein (CRP), TNF receptor 2, and IL-6 (Kurowska et al. 2023; Montonen et al. 2013), and impart a positive effect on depression and anxiety, leading to an improvement in their score (Kurowska et al. 2023). Hence, incorporating whole grains into dietary interventions may play a complementary role in the treatment of depression.

11.6 Conclusion

The impact of fibers on MS varies widely, encompassing benefits such as delaying the onset of neuroinflammation, reducing symptom severity and MS incidence, and enhancing cognitive function while improving depression and anxiety. The positive effect of fibers such as guar gum and cellulose on the onset and prognosis of MS in the EAE mice model may potentially translate to humans through the adoption of a whole-grain diet. Nevertheless, further research is needed to compare the impact of these two individual agents, especially at higher concentrations, considering that diets low in whole grains have been associated with a higher prevalence of MS.

11.7 Summary

For decades, disease-modifying therapy has been the cornerstone of MS management, focusing on motor function restoration, disease relapse prevention, and remission maintenance. However, optimizing lifestyle and diet remains crucial for coping with MS-related fatigue and psychological comorbidities. This chapter delves into the multifaceted impact of dietary interventions, emphasizing the role of whole grains and fibers in MS management. Experimental models highlight the potential of fibers like guar gum and cellulose in alleviating MS symptoms by modulating gut microbiota and inflammatory markers. Whole grains offer broader health benefits, including improved cognitive function and reduced risks of various diseases. Integrating whole grains and specific fibers into the diet may delay neuroinflammation onset, reduce symptom severity, and enhance overall well-being for MS patients. Further research is needed to solidify these findings and optimize dietary recommendations for MS management.

References

Berer K, Martínez I, Walker A, Kunkel B, Schmitt-Kopplin P, Walter J, Krishnamoorthy G (2018) Dietary non-fermentable fiber prevents autoimmune neurological disease by changing gut metabolic and immune status. Sci Rep 8(1):10431. https://doi.org/10.1038/s41598-018-28839-3

Callejo M, Mondejar-Parreño G, Barreira B, Izquierdo-Garcia JL, Morales-Cano D, Esquivel-Ruiz S, Moreno L, Cogolludo Á, Duarte J, Perez-Vizcaino F (2018) Pulmonary arterial hypertension affects the rat gut microbiome. Sci Rep 8(1):9681. https://doi.org/10.1038/s41598-018-27682-w

Chan R, Chan D, Woo J (2013) A cross sectional study to examine the association between dietary patterns and cognitive impairment in older Chinese people in Hong Kong. J Nutr Health Aging 17(9):757–765. https://doi.org/10.1007/s12603-013-0348-5

Connolly ML, Lovegrove JA, Tuohy KM (2010) In vitro evaluation of the microbiota modulation abilities of different sized whole oat grain flakes. Anaerobe 16(5):483–488. https://doi.org/10.1016/j.anaerobe.2010.07.001

Evers I, Heerings M, de Roos NM, Jongen PJ, Visser LH (2022) Adherence to dietary guidelines is associated with better physical and mental quality of life: results from a cross-sectional survey among 728 Dutch MS patients. Nutr Neurosci 25(8):1633–1640. https://doi.org/10.1080/1028415X.2021.1885240

Fettig NM, Robinson HG, Allanach JR, Davis KM, Simister RL, Wang EJ, Sharon AJ, Ye J, Popple SJ, Seo JH, Gibson DL, Crowe SA, Horwitz MS, Osborne LC (2022) Inhibition of Th1 activation and differentiation by dietary guar gum ameliorates experimental autoimmune encephalomyelitis. Cell Rep 40(11):111328. https://doi.org/10.1016/j.celrep.2022.111328

Fitzgerald KC, Tyry T, Salter A, Cofield SS, Cutter G, Fox R, Marrie RA (2018) Diet quality is associated with disability and symptom severity in multiple sclerosis. Neurology 90(1):e1–e11. https://doi.org/10.1212/wnl.0000000000004768

Goufo P, Trindade H (2017) Factors influencing antioxidant compounds in rice. Crit Rev Food Sci Nutr 57(5):893–922. https://doi.org/10.1080/10408398.2014.922046

Horii Y, Uchiyama K, Toyokawa Y, Hotta Y, Tanaka M, Yasukawa Z, Tokunaga M, Okubo T, Mizushima K, Higashimura Y, Dohi O, Okayama T, Yoshida N, Katada K, Kamada K, Handa O, Ishikawa T, Takagi T, Konishi H, Itoh Y (2016) Partially hydrolyzed guar gum enhances colonic epithelial wound healing via activation of RhoA and ERK1/2. Food Funct 7(7):3176–3183. https://doi.org/10.1039/c6fo00177g

Hung TV, Suzuki T (2016) Dietary fermentable fiber reduces intestinal barrier defects and inflammation in colitic mice. J Nutr 146(10):1970–1979. https://doi.org/10.3945/jn.116.232538

Jahromi SR, Toghae M, Jahromi MJR, Aloosh M (2012) Dietary pattern and risk of multiple sclerosis. Iran J Neurol 11(2):47

Kadowaki A, Saga R, Lin Y, Sato W, Yamamura T (2019) Gut microbiota-dependent CCR9+CD4+ T cells are altered in secondary progressive multiple sclerosis. Brain 142(4):916–931. https://doi.org/10.1093/brain/awz012

Kim S, Rigatto K, Gazzana MB, Knorst MM, Richards EM, Pepine CJ, Raizada MK (2020) Altered gut microbiome profile in patients with pulmonary arterial hypertension. Hypertension 75(4):1063–1071. https://doi.org/10.1161/HYPERTENSIONAHA.119.14294

Kurowska A, Ziemichód W, Herbet M, Piątkowska-Chmiel I (2023) The role of diet as a modulator of the inflammatory process in the neurological diseases. Nutrients 15(6):1436. https://www.mdpi.com/2072-6643/15/6/1436

Lalan S, Khan M, Schlakman B, Penman A, Gatlin J, Herndon R (2012) Differentiation of neuromyelitis optica from multiple sclerosis on spinal magnetic resonance imaging. Int J MS Care 14(4):209–214. https://doi.org/10.7224/1537-2073-14.4.209

Montonen J, Boeing H, Fritsche A, Schleicher E, Joost H-G, Schulze MB, Steffen A, Pischon T (2013) Consumption of red meat and whole-grain bread in relation to biomarkers of obesity, inflammation, glucose metabolism and oxidative stress. Eur J Nutr 52(1):337–345. https://doi.org/10.1007/s00394-012-0340-6

Mudgil D, Barak S, Khatkar BS (2014) Guar gum: processing, properties and food applications-a review. J Food Sci Technol 51(3):409–418. https://doi.org/10.1007/s13197-011-0522-x

Naito Y, Takagi T, Katada K, Uchiyama K, Kuroda M, Kokura S, Ichikawa H, Watabe J, Yoshida N, Okanoue T, Yoshikawa T (2006) Partially hydrolyzed guar gum down-regulates colonic inflammatory response in dextran sulfate sodium-induced colitis in mice. J Nutr Biochem 17(6):402–409. https://doi.org/10.1016/j.jnutbio.2005.08.010

Nirmala Prasadi VP, Joye IJ (2020) Dietary fibre from whole grains and their benefits on metabolic health. Nutrients 12(10):3045. https://doi.org/10.3390/nu12103045

Ochoa-Repáraz J, Kirby TO, Kasper LH (2018) The gut microbiome and multiple sclerosis. Cold Spring Harb Perspect Med 8(6):a029017

Rajaram S, Jones J, Lee GJ (2019) Plant-based dietary patterns, plant foods, and age-related cognitive decline. Adv Nutr 10:S422–S436. https://doi.org/10.1093/advances/nmz081

Rezaeimanesh N, Ariyanfar S, Sahraian MA, Naser Moghadasi A, Ghorbani Z, Razegh-Jahromi S (2021) Whole grains and legumes consumption in association with neuromyelitis optica spectrum disorder odds. Curr J Neurol 20(3):131–138. https://doi.org/10.18502/cjn.v20i3.7688

Sanada TJ, Hosomi K, Park J, Naito A, Sakao S, Tanabe N, Kunisawa J, Tatsumi K, Suzuki T (2023) Partially hydrolyzed guar gum suppresses the progression of pulmonary arterial hypertension in a SU5416/hypoxia rat model. Pulm Circ 13(3):e12266. https://doi.org/10.1002/pul2.12266

Saresella M, Mendozzi L, Rossi V, Mazzali F, Piancone F, LaRosa F, Marventano I, Caputo D, Felis GE, Clerici M (2017) Immunological and clinical effect of diet modulation of the gut microbiome in multiple sclerosis patients: a pilot study. Front Immunol 8:1391. https://doi.org/10.3389/fimmu.2017.01391

Sørensen TL, Tani M, Jensen J, Pierce V, Lucchinetti C, Folcik VA, Qin S, Rottman J, Sellebjerg F, Strieter RM, Frederiksen JL, Ransohoff RM (1999) Expression of specific chemokines and chemokine receptors in the central nervous system of multiple sclerosis patients. J Clin Invest 103(6):807–815. https://doi.org/10.1172/jci5150

Sørensen TL, Trebst C, Kivisäkk P, Klaege KL, Majmudar A, Ravid R, Lassmann H, Olsen DB, Strieter RM, Ransohoff RM, Sellebjerg F (2002) Multiple sclerosis: a study of CXCL10 and CXCR3 co-localization in the inflamed central nervous system. J Neuroimmunol 127(1-2):59–68. https://doi.org/10.1016/s0165-5728(02)00097-8

Takagi T, Naito Y, Higashimura Y, Uchiroda C, Mizushima K, Ohashi Y, Yasukawa Z, Ozeki M, Tokunaga M, Okubo T, Katada K, Kamada K, Uchiyama K, Handa O, Itoh Y, Yoshikawa T (2016) Partially hydrolysed guar gum ameliorates murine intestinal inflammation in association with modulating luminal microbiota and SCFA. Br J Nutr 116(7):1199–1205. https://doi.org/10.1017/S0007114516003068

Chapter 12
Diet and Its Potential Impact on the Prognosis of Multiple Sclerosis: Fasting Diets

Amna Baig, Haia M. R. Abdulsamad, Nadia Rabeh, Sara Aljoudi, Zakia Dimassi, and Hamdan Hamdan

Abstract This chapter explores the therapeutic potential of fasting diets in multiple sclerosis (MS), a chronic neurodegenerative disease marked by inflammation and neuronal damage. Fasting regimens like intermittent fasting, time-restricted feeding, and the fast-mimicking diet are examined for their ability to modulate MS pathogenesis. Emphasis is placed on their anti-inflammatory effects and impact on key pathways, particularly the mammalian target of the rapamycin (mTOR) pathway, as well as the pathways through which they influence gut microbiota. While clinical trials demonstrate safety and positive outcomes in MS patients, optimizing fasting parameters and long-term effects necessitate further exploration. Large-scale trials comparing fasting types, associated benefits and risks, and considering diverse patient demographics are essential for establishing universal applicability as a complementary dietary approach to pharmaceutical MS treatments.

Keywords Multiple sclerosis (MS) · Fasting diets · Intermittent fasting · Calorie restriction · Inflammation

---

A. Baig · H. M. R. Abdulsamad · N. Rabeh · S. Aljoudi · H. Hamdan (✉)
Department of Biological Sciences, College of Medicine and Health Sciences, Khalifa University, Abu Dhabi, United Arab Emirates
e-mail: hamdan.hamdan@ku.ac.ae

Z. Dimassi
Department of Medical Sciences, College of Medicine and Health Sciences, Khalifa University, Abu Dhabi, United Arab Emirates

© The Author(s), under exclusive license to Springer Nature Singapore Pte Ltd. 2024
H. Hamdan (ed.), *Exploring the Effects of Diet on the Development and Prognosis of Multiple Sclerosis (MS)*, Nutritional Neurosciences,
https://doi.org/10.1007/978-981-97-4673-6_12

Abbreviations

ADF	Alternate day fasting
BMI	Body mass index
CNS	Central nervous system
CR	Calorie restriction
EAE	Experimental autoimmune encephalomyelitis
FMD	Fast-mimicking diet
IF	Intermittent fasting
IGF-1	Insulin-like growth factor-1
IL	Interleukin
MS	Multiple sclerosis
mTOR	Mammalian target of the rapamycin
RRMS	Relapse remitting MS
Th	T helper
TRF	Time-restricted feeding

Learning Objectives
- Describe the potential therapeutic benefits of various fasting diets, including intermittent fasting (IF), time-restricted feeding (TRF), and the fast-mimicking diet (FMD), in managing MS.
- Identify the anti-inflammatory effects of fasting diets and their impact on key pathways such as the mammalian target of rapamycin (mTOR) pathway.
- Explain how fasting diets influence gut microbiota and contribute to the modulation of immune responses in MS patients.
- Assess the findings from clinical trials on the safety, feasibility, and potential benefits of fasting diets in MS patients.
- Discuss the future research directions needed to optimize fasting parameters, compare different fasting types, and establish their long-term effects and universal applicability in MS management.

12.1 Introduction

Multiple sclerosis (MS) is a chronic neurodegenerative disease of the central nervous system (CNS) that causes inflammation, demyelination, and axonal damage to neurons. MS is a multifactorial disease, with genetic, immunologic, and environmental factors contributing to the disease's onset and evolution (Felicetti et al. 2022). The underlying pathogenesis, mediated by CD4+ T helper (Th) 17 and Th1 cell subtypes, involves activating inflammatory markers and impairing the immune system, leading to neurodegenerative effects affecting neurons within the CNS (Choi et al. 2016). MS patients endure various disease manifestations, such as pain, weakness, fatigue, anxiety, and depression (Hassan et al. 2021).

A growing body of evidence suggests that dietary components and regimens are associated with the progression, relapse rate, and development of MS (Stoiloudis et al. 2022). One regimen showing promising effects is fasting. The term "fasting" is used here to encompass the many different types of fasting used for research thus far, including, but not limited to, pure calorie restriction, intermittent fasting, and the fast-mimicking diet (Visioli et al. 2022). Within these categories, there is variation in the number of calories consumed, the breakdown of the composition of those calories, the hours spent fasting, and whether fasting is abstaining from all foods or simply keeping one's calorie intake below a certain number during the 'fasting hours.' This chapter will discuss different fasting diets, the research that looks into the benefits and drawbacks of implementing fasting diets in animal models, and the implications of utilizing these diets for patients with MS.

12.2 Fasting Diets and Their Mechanisms of Inflammatory Modulation

It has been well-established that fasting leads to a systemic reduction of inflammatory markers, with previous research heavily focusing on the role of fasting in promoting weight loss (Mulas et al. 2023). Fasting diets inherently rely on a degree of calorie restriction (CR) and have been shown to cause potent anti-inflammatory effects (Morales-Suarez-Varela et al. 2021) and promote resistance against cognitive decline (Gudden et al. 2021). Amongst the various forms of fasting, Intermittent fasting (IF), time-restricted feeding (TRF), and fast-mimicking diet (FMD) are the three main subtypes that will be discussed in this chapter.

IF is a diet that involves cycles of reduced or no calorie consumption followed by regular unrestricted eating periods (Tang et al. 2023). A subcategory of IF is alternate day fasting (ADF), characterized by a day of fasting where individuals consume 25% of their regular intake, followed by a day of regular, unrestricted eating (Cui et al. 2020). When the fasting period is sustained for an extended period, the intermittent CR prompts the metabolism to transition from glucose utilization to lipolysis, the breakdown of fat. This shift triggers a series of metabolic, cellular, and circadian alterations. In both animal and human studies, this shift may support the improvement of many different diseases, specifically neurological diseases such as MS. IF has been linked to lowering the risk of neurological diseases due to its capacity to induce structural and functional changes in the brain, as indicated by Gudden et al.'s research, thus positioning IF as a promising intervention in both animals and humans for improving symptoms and slowing disease progression. One of the potential pathways through which IF affects the pathogenesis of MS is mediated by its effect on the Mammalian target of the rapamycin (mTOR) pathway. Under normal conditions, mTOR regulates cellular proliferation and autophagy. The mTOR pathway is downregulated during periods of fasting, thus reducing protein synthesis and decreasing the accumulation of free radicals (Johnson et al. 2013).

TRF is a form of fasting where food consumption is limited to certain hours within the day. Fasting periods usually range from 4–12 h, followed by a period of eating without CR. The fasting and feeding periods are tailored to align with the circadian rhythm, such that fasting occurs during stages of rest, while feeding occurs during awakened, active states (Soliman 2022). TRF is an effective method of reducing weight loss as it is deemed more flexible than other fasting regimens (Soliman 2022). Research also shows that TRF can be a beneficial add-on for cancer treatment, decrease oxidative stress markers, and improve insulin sensitivity (Soliman 2022). However, clinical trials investigating TRF's ability to reduce inflammatory markers such as tumor necrosis factor (TNF), C-reactive protein, and interleukin (IL)-6 found no significant improvement (Cienfuegos et al. 2020; Sutton et al. 2018).

The FMD involves fasting cycles with restricted calorie intake and, if desired, emphasizing certain food groups. Individuals do not abstain from eating but engage in CR for several consecutive days during the week, followed by days of regular, unrestricted eating for the remainder of the week. Improvements in cellular regeneration and protection when adhering to the FMD are attributed to reduced levels of insulin-like growth factor-1 (IGF-1), as IGF-1 is an inhibitor of these processes (Choi et al. 2017). Research on FMD also found that it plays a notable role in immune system regulation (Tang et al. 2023). Some of the effects FMD can induce include decreasing autoimmunity, promoting the circulation of lymphocytes, activating antitumor activity, and increasing anti-inflammatory factors (Choi et al. 2016, 2017; Cortellino et al. 2023). Of these downstream effects, reductions in autoimmunity are of prime importance as reactive immune cells are thought to be the driving factor in the pathogenesis of MS (Afshar et al. 2019).

Although fasting diets are deemed generally safe, they are not devoid of side effects or contraindications (Fay-Watt et al. 2023; Pavlou et al. 2023). Fasting regimens are not advisable for individuals suffering from eating disorders, as the diet may trigger or exacerbate their symptoms (Grajower and Horne 2019). A common side effect of fasting diets is the increased onset of headaches secondary to dehydration (Varady et al. 2022). Adequate macro- and micronutrient intake during periods of food consumption is essential to avoid developing malnutrition, with potential side effects including dizziness, syncope, nausea, and weakness. Moreover, individuals with comorbidities, such as diabetes mellitus, and individuals who are on insulin or sulfonylureas may be at a greater risk of becoming hypoglycemic if they engage in fasting diets, thus requiring extra care (Grajower and Horne 2019).

12.3 Fasting and Multiple Sclerosis

The available literature had previously focused on reducing the number of calories in the diet, which had shown positive effects on the rate of MS development. For example, in a study with Lewis rats, the experimental group was subjected to a severely restricted diet (66% restriction compared to the controls) and followed to see if they developed experimental autoimmune encephalomyelitis (EAE), an animal model of MS, after immunization with Freund's adjuvant, which is designed to induce EAE. The rats on the CR diet were protected against EAE development (Esquifino et al. 2004, 2007). However, it is unfeasible to subject human patients to a 66% CR as in the study above, so another study attempted to see the results of a

40% CR. The experimental mice were given the 40% restricted diet for 5 weeks before having EAE induced. This group was compared to a control group and a group fed a high-fat, high-calorie diet. The restricted calorie diet scored higher on the clinical score of EAE, meaning there was less demyelination, inflammation, and axonal damage (Piccio et al. 2008). Although the studies reported positive results, it is not feasible to introduce extensive CR for long periods in patients with MS. The alternative is introducing the FMD, which restricts calories based on a cycle.

12.4 Fast-Mimicking Diet

Choi et al. and Bai et al. recently conducted clinical studies using EAE-induced mice models to examine the effect of FMD on the pathogenesis of MS. These studies introduced 3 days of restricted calorie intake (with low protein, low carbohydrates, and emphasis on high fiber content), followed by 4 days of eating without CR. The mice were assessed for clinical severity (symptomatic score), white blood cell infiltration in the CNS, and demyelination extent. FMD was shown to have beneficial effects on the clinical severity of MS and, in a small percentage, eradicated the symptoms of MS. In addition, there was a marked improvement in the extent of CNS inflammation with less immune cell infiltration. These studies also found that FMD promoted oligodendrocyte regeneration and elevated remyelination factors. T-cell-mediated demyelination and neurodegeneration are classic hallmarks of MS, and both Choi et al. and Bai et al. found that FMD has mitigating effects on these factors (Bai et al. 2021; Choi et al. 2016). Both studies reported no apparent drawbacks to the FMD in the EAE-induced mice.

Despite the promising results from EAE-induced mouse models, clinical trials investigating the efficacy of FMD in MS patients still need to be completed. Choi et al. included a randomized pilot trial to explore whether diet benefits MS patients. They recruited sixty relapsing-remitting patients with MS and assigned them to three groups: FMD followed by either a Mediterranean diet, a ketogenic diet, or a control diet (Choi et al. 2016). The results showed that the FMD patients developed fewer severe adverse events of MS, such as urinary tract infections, than the control patients. They concluded that FMD is safe and feasible, but further investigation is required. A more comprehensive study should evaluate the breakdown of the FMD that offers optimal results. Additional information is needed to determine the ideal number of calories and dietary composition (i.e., less protein-heavy, more fiber-heavy, more plant-based) for the calorie-restrictive days and whether these factors impact how effective FMD is for MS (Bai et al. 2021; Choi et al. 2016).

Fitzgerald et al. conducted a pilot randomized control study with 36 MS patients to compare the effects on weight loss and the feasibility of two different types of fasting diets, along with a control diet. They randomized patients to either a pure CR diet, with patients eating a 22% reduction in calorie intake, or to an intermittent CR diet, with patients having a 75% reduction in calorie intake for 2 days of the week, with regular, unrestricted eating in the other five (Fitzgerald et al. 2022). Healthy weight loss in patients with MS has been associated with increased insulin sensitivity, reduced markers of inflammation, and the potential benefits of fasting aside from weight loss. The study found that both CR/fasting diets were associated with weight loss and, interestingly enough, higher emotional well-being scores (Fitzgerald et al. 2022). Currently, there is an ongoing single-arm, open-label clinical trial assessing the safety and tolerability of FMD in patients with relapse-remitting MS (RRMS) (NCT06039007). In addition to their pharmacological regimens, the participants are instructed to complete three cycles of FMD every 60 days. The prescribed 7-day FMD involves 1100 kcal on the first day, followed by 800 kcal for the remainder of the week.

12.5 Intermittent Fasting

IF appears to have a favorable impact on MS, potentially through its influence on gut microbiota and immune responses, as indicated by both animal and human studies (Gudden et al. 2021). Cignarella et al. found that IF alters the gut microbiota's composition. Increased presence of the Bacteroidaceae, Lactobacillaceae, and Prevotellaceae families contributed to increased antioxidative effects (Cignarella et al. 2018). The increased abundance of lactobacilli resulting from IF has positive effects, including the reduction of inflammatory immune responses, which can be beneficial for individuals with MS, according to research by Morales-Suarez-Varela et al. (2021). IF also induced reductions in inflammatory markers such as IL-17, thus leading to less T-cell activation. Fitzgerald et al. conducted a small randomized controlled trial on patients with MS. They found that patients on CR had significantly fewer Th1 cells, one of the many hallmarks of MS inflammation (Fitzgerald et al. 2022).

There has been some debate about whether introducing IF in the early stages of MS may lead to a worse prognosis. Razeghi et al. used the EAE mice model to demonstrate precise mechanisms through which IF can positively impact brain-related disease models, regardless of when the dietary intervention is initiated. In this study, once EAE was induced in mice, they were split into two groups, one where IF was implemented at an early stage and another where IF was initiated at a later stage (Razeghi et al. 2016). Three factors were evaluated: clinical severity score, histological evaluation to assess the extent of demyelination, and lymphocyte proliferation assay to assess inflammatory marker levels. There was no evidence that fasting had unfavorable effects on the disease course of the EAE, regardless of when the diet was introduced. Instead, the earlier the diet was started, the less clinically severe the progression of EAE, with mice showing less demyelination, better clinical scores, and improved suppression of autoimmune inflammatory responses. Razeghi et al.'s study provides some weight to the argument that fasting during the early stage of MS is more effective than starting the diet later in the disease progression (Razeghi et al. 2016).

As mentioned earlier, IF causes a change in the gut microbiota, which leads to ameliorating effects on the progression of MS. Cignarella et al. transplanted the gut microbiota from IF mice to non-IF mice with MS and found that they had a reduction in clinically severe symptoms after the transplant, despite not being on the IF diet themselves. Cignarella et al. then conducted a small RCT involving five MS patients and nine control subjects. The participants were instructed to adhere to an ADF regimen for 15 days, resulting in gut microbiota changes like those observed in mice (Cignarella et al. 2018).

12.6 Timed-Restricted Feeding

Of the fasting interventions, individuals seem to adhere to the TRF diet more than other fasting diets (Roman et al. 2020). A randomized control trial assessing the safety and tolerability of TRF in patients with RRMS reported high adherence rates. However, the measurement was limited to self-reported calorie intake reports, which the study participants inconsistently submitted. Patients were asked to fast for 16 h each week for 2 days, followed by an 8-h eating window. The intervention was implemented for 6 months, and patients had to report their caloric intake on their fasting days. No adverse events were noted during the intervention (Roman et al. 2020).

Wingo et al. conducted a pilot study in which patients with MS were put on a TRF intervention to determine adherence rates and tolerance to the diet. This study used the 16-hour fast and 8-hour window of eating method for 8 weeks. There were no limitations on the food consumed other than the request that patients adhere to their regular diet, with the only change being the window in which they were allowed to eat. Overall, the study concluded that this form of fasting was feasible for patients with MS, as there was a high retention of participants, positive feedback on the diet and experience, and a high level of self-reported adherence to the diet. Although this study only assessed feasibility and adherence, patients reported qualitative improvements in quality-of-life scores, such as better sleep, less fatigue, and, most importantly, less pain (Wingo et al. 2023).

12.7 Future Outlook

This chapter's research highlights fasting's therapeutic potential in MS treatment. Evidence from the clinical trials demonstrates the safety and tolerability of various fasting diets. Before such regimens can be transferred to clinical practice, additional research is needed to understand the mechanisms underlying their effect. Future research should investigate whether the protective effects of fasting diets apply universally, regardless of factors like age, body mass index (BMI), total caloric intake, and the timing and composition of nutrient intake. Additionally, while there is theoretical potential for these fasting diets to benefit the treatment of MS, there remains a need to conduct and gather additional experimental data. Further research is required to explore the potential benefits of fasting in the context of neurocognitive diseases.

The optimal frequency and duration of fasting still need to be determined. The clinical trials presented in this chapter mainly implemented fasting regimens for only 6 months, which entails a need for longitudinal trials to understand the long-term effects of fasting diets on MS. Longitudinal trials are also valuable for understanding the impact of fasting during different stages of the disease. In addition, there is an urgent need to define what type of fasting diet is the most beneficial, most applicable, and most feasible for MS patients. Thus far, most of the research has established that different forms of fasting are helpful for the prognosis of MS, but it is now time to compare these different types against each other on a large scale. If large-scale human trials of fasting diets prove to have successful outcomes as they have done in animal models, fasting diets could become a reliable addition to current pharmacological treatment for MS.

12.8 Conclusion

In conclusion, MS is a complex neurodegenerative disease involving genetic, immunologic, and environmental factors, with inflammation and T cell dysregulation playing a crucial role. This chapter explores fasting diets as potential modulatory strategies for MS. Fasting shows promise in reducing inflammatory markers, with IF demonstrating positive effects on neurobiological health. The FMD, mainly studied in EAE-induced mouse models, exhibits promise in improving clinical severity, inflammation, and demyelination in MS patients. Clinical trials demonstrate safety, but optimization of FMD parameters is necessary. Research on fasting's effects on gut microbiota and immune responses provides valuable insights. While encouraging, more extensive and longitudinal trials are needed to establish the long-term effects of fasting on MS. Comparisons between fasting types and exploration of universal applicability across demographics are crucial. If validated, fasting diets could be valuable to current MS treatments.

12.9 Summary

This chapter explores the therapeutic potential of fasting diets in managing MS, a chronic neurodegenerative disease characterized by inflammation and neuronal damage. Various fasting regimens, including intermittent fasting, time-restricted feeding, and the fast-mimicking diet, are examined for their ability to modulate MS pathogenesis. Emphasis is placed on their anti-inflammatory effects and impact on key pathways, particularly the mammalian target of the mTOR pathway, and their influence on gut microbiota. Clinical trials demonstrate the safety and positive outcomes of fasting in MS patients, highlighting the need for further research to optimize fasting parameters and understand long-term effects. Large-scale trials comparing different fasting types and considering diverse patient demographics are essential for establishing fasting as a complementary dietary approach to pharmaceutical MS treatments.

References

Afshar B, Khalifehzadeh-Esfahani Z, Seyfizadeh N, Rezaei Danbaran G, Hemmatzadeh M, Mohammadi H (2019) The role of immune regulatory molecules in multiple sclerosis. J Neuroimmunol 337:577061. https://doi.org/10.1016/j.jneuroim.2019.577061

Bai M, Wang Y, Han R, Xu L, Huang M, Zhao J, Lin Y, Song S, Chen Y (2021) Intermittent caloric restriction with a modified fasting-mimicking diet ameliorates autoimmunity and promotes recovery in a mouse model of multiple sclerosis. J Nutr Biochem 87:108493

Choi IY, Piccio L, Childress P, Bollman B, Ghosh A, Brandhorst S, Suarez J, Michalsen A, Cross AH, Morgan TE, Wei M, Paul F, Bock M, Longo VD (2016) A diet mimicking fasting promotes regeneration and reduces autoimmunity and multiple sclerosis symptoms. Cell Rep 15(10):2136–2146. https://doi.org/10.1016/j.celrep.2016.05.009

Choi IY, Lee C, Longo VD (2017) Nutrition and fasting mimicking diets in the prevention and treatment of autoimmune diseases and immunosenescence. Mol Cell Endocrinol 455:4–12. https://doi.org/10.1016/j.mce.2017.01.042

Cienfuegos S, Gabel K, Kalam F, Ezpeleta M, Wiseman E, Pavlou V, Lin S, Oliveira ML, Varady KA (2020) Effects of 4- and 6-h time-restricted feeding on weight and cardiometabolic health: a randomized controlled trial in adults with obesity. Cell Metab 32(3):366–378. https://doi.org/10.1016/j.cmet.2020.06.018

Cignarella F, Cantoni C, Ghezzi L, Salter A, Dorsett Y, Chen L, Phillips D, Weinstock GM, Fontana L, Cross AH (2018) Intermittent fasting confers protection in CNS autoimmunity by altering the gut microbiota. Cell Metab 27(6):1222–1235

Cortellino S, Quagliariello V, Delfanti G, Blaževitš O, Chiodoni C, Maurea N, Di Mauro A, Tatangelo F, Pisati F, Shmahala A, Lazzeri S, Spagnolo V, Visco E, Tripodo C, Casorati G, Dellabona P, Longo VD (2023) Fasting mimicking diet in mice delays cancer growth and reduces immunotherapy-associated cardiovascular and systemic side effects. Nat Commun 14(1):5529. https://doi.org/10.1038/s41467-023-41066-3

Cui Y, Cai T, Zhou Z, Mu Y, Lu Y, Gao Z, Wu J, Zhang Y (2020) Health effects of alternate-day fasting in adults: a systematic review and meta-analysis. Front Nutr 7:586036. https://doi.org/10.3389/fnut.2020.586036

Esquifino A, Cano P, Jimenez V, Cutrera R, Cardinali D (2004) Experimental allergic encephalomyelitis in male Lewis rats subjected to calorie restriction. J Physiol Biochem 60(4):245–252

Esquifino AI, Cano P, Jimenez-Ortega V, Fernández-Mateos MP, Cardinali DP (2007) Immune response after experimental allergic encephalomyelitis in rats subjected to calorie restriction. J Neuroinflammation 4:1–10

Fay-Watt V, O'Connor S, Roshan D, Romeo AC, Longo VD, Sullivan FJ (2023) The impact of a fasting mimicking diet on the metabolic health of a prospective cohort of patients with prostate cancer: a pilot implementation study. Prostate Cancer Prostatic Dis 26(2):317–322. https://doi.org/10.1038/s41391-022-00528-3

Felicetti F, Tommasin S, Petracca M, De Giglio L, Gurreri F, Ianniello A, Nistri R, Pozzilli C, Ruggieri S (2022) Eating hubs in multiple sclerosis: exploring the relationship between mediterranean diet and disability status in Italy. Front Nutr 9:882426. https://doi.org/10.3389/fnut.2022.882426

Fitzgerald KC, Bhargava P, Smith MD, Vizthum D, Henry-Barron B, Kornberg MD, Cassard SD, Kapogiannis D, Sullivan P, Baer DJ (2022) Intermittent calorie restriction alters T cell subsets and metabolic markers in people with multiple sclerosis. EBioMedicine 82:104124

Grajower MM, Horne BD (2019) Clinical management of intermittent fasting in patients with diabetes mellitus. Nutrients 11(4):873. https://doi.org/10.3390/nu11040873

Gudden J, Arias Vasquez A, Bloemendaal M (2021) The effects of intermittent fasting on brain and cognitive function. Nutrients 13(9):3166

Hassan A, Merghany N, Ouchkat F, Regragui W, Kedah H, Hamdy SM, Kishk NA (2021) Impact of Ramadan fasting on disease activity in patients with multiple sclerosis: a multicenter study. Nutr Neurosci 1–10

Johnson SC, Rabinovitch PS, Kaeberlein M (2013) mTOR is a key modulator of ageing and age-related disease. Nature 493(7432):338–345

Morales-Suarez-Varela M, Collado Sánchez E, Peraita-Costa I, Llopis-Morales A, Soriano JM (2021) Intermittent fasting and the possible benefits in obesity, diabetes, and multiple sclerosis: a systematic review of randomized clinical trials. Nutrients 13(9):3179

Mulas A, Cienfuegos S, Ezpeleta M, Lin S, Pavlou V, Varady KA (2023) Effect of intermittent fasting on circulating inflammatory markers in obesity: a review of human trials. Front Nutr 10:543

Pavlou V, Cienfuegos S, Lin S, Ezpeleta M, Ready K, Corapi S, Wu J, Lopez J, Gabel K, Tussing-Humphreys L, Oddo VM, Alexandria SJ, Sanchez J, Unterman T, Chow LS, Vidmar AP, Varady KA (2023) Effect of time-restricted eating on weight loss in adults with type 2 diabetes: a randomized clinical trial. JAMA Netw Open 6(10):e2339337. https://doi.org/10.1001/jamanetworkopen.2023.39337

Piccio L, Stark JL, Cross AH (2008) Chronic calorie restriction attenuates experimental autoimmune encephalomyelitis. J Leucocyte Biol 84(4):940–948

Razeghi JS, Ghaemi A, Alizadeh A, Sabetghadam F, Moradi TH, Togha M (2016) Effects of intermittent fasting on experimental autoimmune encephalomyelitis in C57BL/6 mice. Iran J Allergy Asthma Immunol 15(3):212–219

Roman SN, Fitzgerald KC, Beier M, Mowry EM (2020) Safety and feasibility of various fasting-mimicking diets among people with multiple sclerosis. Mult Scler Relat Disord 42:102149. https://doi.org/10.1016/j.msard.2020.102149

Soliman GA (2022) Intermittent fasting and time-restricted eating role in dietary interventions and precision nutrition. Front Public Health 10:1017254. https://doi.org/10.3389/fpubh.2022.1017254

Stoiloudis P, Kesidou E, Bakirtzis C, Sintila S-A, Konstantinidou N, Boziki M, Grigoriadis N (2022) The role of diet and interventions on multiple sclerosis: a review. Nutrients 14(6):1150

Sutton EF, Beyl R, Early KS, Cefalu WT, Ravussin E, Peterson CM (2018) Early time-restricted feeding improves insulin sensitivity, blood pressure, and oxidative stress even without weight loss in men with prediabetes. Cell Metab 27(6):1212–1221. https://doi.org/10.1016/j.cmet.2018.04.010

Tang D, Tang Q, Huang W, Zhang Y, Tian Y, Fu X (2023) Fasting: from physiology to pathology. Adv Sci 10(9):2204487. https://doi.org/10.1002/advs.202204487

Varady KA, Cienfuegos S, Ezpeleta M, Gabel K (2022) Clinical application of intermittent fasting for weight loss: progress and future directions. Nat Rev Endocrinol 18(5):309–321. https://doi.org/10.1038/s41574-022-00638-x

Visioli F, Mucignat-Caretta C, Anile F, Panaite S-A (2022) Traditional and medical applications of fasting. Nutrients 14(3):433

Wingo BC, Rinker JR, Green K, Peterson CM (2023) Feasibility and acceptability of time-restricted eating in a group of adults with multiple sclerosis. Front Neurol, 13,1087126

Chapter 13
Diet and Its Potential Impact on the Prognosis of Multiple Sclerosis: Mediterranean Diet

Amna Baig, Haia M. R. Abdulsamad, Sara Aljoudi, Nadia Rabeh, Zakia Dimassi, and Hamdan Hamdan

Abstract The Mediterranean diet (MD) is touted to influence the development of MS. This chapter will expand on the suggested roles of the different components within the Mediterranean diet on the development and progression of MS. It will explore how these dietary components and regimens influence the mechanistic pathways associated with MS. The MD, emphasizing fruits, vegetables, and low saturated fats, shows promise in alleviating MS symptoms. Increased consumption of fruits and vegetables has been associated with a less severe form of MS attributed to antioxidants. Conversely, limited saturated fats align with a better prognosis, as they are involved in activating pro-inflammatory toll-like receptors, which are linked to a poor MS prognosis. Polyunsaturated and monounsaturated fats in the MD exhibit protective effects. Human trials examining the role of the MD in MS have been thus far limited, emphasizing the need for extensive research to validate the MD's therapeutic benefits in managing MS symptoms and long-term disability.

Keywords Multiple sclerosis (MS) · Mediterranean diet · Oxidative stress · Polyunsaturated fatty acids · Inflammation

---

A. Baig · H. M. R. Abdulsamad · S. Aljoudi · N. Rabeh H. Hamdan (✉)
Department of Biological Sciences, College of Medicine and Health Sciences, Khalifa University, Abu Dhabi, United Arab Emirates
e-mail: hamdan.hamdan@ku.ac.ae

Z. Dimassi
Department of Medical Sciences, College of Medicine and Health Sciences, Khalifa University, Abu Dhabi, United Arab Emirates

© The Author(s), under exclusive license to Springer Nature Singapore Pte Ltd. 2024
H. Hamdan (ed.), *Exploring the Effects of Diet on the Development and Prognosis of Multiple Sclerosis (MS)*, Nutritional Neurosciences, https://doi.org/10.1007/978-981-97-4673-6_13

Abbreviations

AhR	Aryl hydrocarbon receptor
CNS	Central nervous system
FSS	Fatigue severity scale
MD	Mediterranean diet
MEDAS	Mediterranean diet assessment tool
MS	Multiple sclerosis
MS-RS	MS-related symptom
MUFA	Monounsaturated fatty acids
PD	Parkinson's disease
PGE1	prostaglandin E1
PGE2	E2
PUFA	Polyunsaturated fatty acids
RRMS	Relapsing-remitting MS
TLR	Toll-like receptors

Learning Objectives
- Identify the key components of the Mediterranean Diet (MD) and their potential benefits for individuals with MS.
- Explain how the dietary components of the MD influence the mechanistic pathways associated with the development and progression of MS.
- Assess the roles of fruits, vegetables, and fats within the MD in reducing oxidative stress and inflammation in MS patients.
- Critically evaluate the findings from human trials examining the efficacy and feasibility of the MD in managing MS symptoms and long-term disability.
- Discuss the limitations of current research on the MD and MS and propose future research directions to validate the diet's therapeutic benefits.

13.1 Introduction

The Mediterranean diet (MD) refers to the general diet consumed by the people of the Mediterranean region. Given the diversity across countries and populations within this region, pinpointing an exact definition specifying the precise intake of calories or grams of macronutrients proves challenging. Food pyramids, as depicted in Fig. 13.1, have been designed that roughly outline the recommended servings for each food group per week (Bach-Faig et al. 2011; Davis et al. 2015).

MD has been shown to have benefits in many neurodegenerative disorders, including Parkinson's disease (PD) and MS (Bisaglia 2022). In fact, long-term adherence to the MD has been linked to protection against the prodromal features of PD, which may manifest decades prior to the official diagnosis (Bisaglia 2022). instead of being strictly restrictive, MD integrates a combination of dietary and

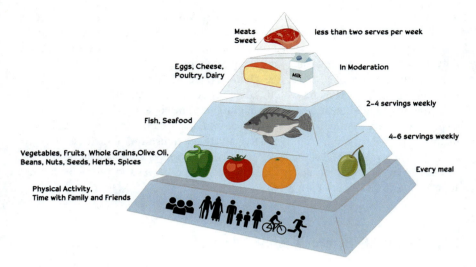

Fig. 13.1 Food pyramid depicting components of a holistic regimen

lifestyle elements (Felicetti et al. 2022). It is distinguished by four primary factors, each contributing to its nutritional philosophy and lifestyle approach that collectively define its positive impact on well-being. First, the bulk of the diet consists of fruits, vegetables, and whole grains, the latter of which will be elaborated upon in another chapter. Second, there is a limited intake of saturated fats with a preference for polyunsaturated and monounsaturated fats, particularly emphasizing the use of olive oil. Finally, as part of the MD, there is a deliberate emphasis on maintaining a limited intake of processed foods (Katz Sand 2018). The subsequent sections will delve into each of these factors individually.

13.2 Multiple Sclerosis and the Consumption of Fruits and Vegetables

Increased intake of fruits and vegetables has been linked with a higher likelihood of experiencing a less clinically severe form of MS in affected individuals (Hadgkiss et al. 2015). The authors of this global study involving participants with MS conducted an investigation on factors such as health-related quality of life, disabilities, and relapse rate alongside a diet history questionnaire to explore potential correlations. The findings revealed that individuals with MS with a higher intake of fruits and vegetables reported an improved quality of life and reduced odds of developing severe clinical symptoms leading to disabilities (Fitzgerald et al. 2018; Hadgkiss et al. 2015). Combining this with the recent systematic review and meta-analysis, which included eight observational studies, it was found that both fruit and vegetable consumption exhibited a protective effect on MS odds (Fotros et al. 2023). Specifically, an increase in fruit consumption by 100 grams per day was associated with a 9% reduction in the odds of developing MS ($I^2 = 0.0\%$, P for heterogeneity = 0.77; pooled OR = 0.91, 95%CI = 0.83, 0.99, P-value = 0.021) (Fotros et al. 2023).

Fruits and vegetables, renowned for containing antioxidants like carotenoids and polyphenols, play a crucial role in reducing oxidative stress, inflammation, and central nervous system (CNS) damage characteristic of MS (Hadgkiss et al. 2015; Hosseini et al. 2018; Rahaman et al. 2023). Cruciferous vegetables containing the essential amino acid tryptophan were highlighted in the study. The metabolites of tryptophan directly influence the peripheral immune system, decreasing its inflammatory response by acting as a ligand to activate the transcription factor aryl hydrocarbon receptor (AhR) (Rothhammer et al. 2016). The systemic activation of AhR, if induced by the appropriate ligand, has been hypothesized to promote immune suppression (Stockinger et al. 2014).

13.3 Multiple Sclerosis and Dietary Fats

Saturated fats, limited in the MD, are associated with a poor MS prognosis due to their role in activating pro-inflammatory toll-like receptors (TLR) (Huang et al. 2012). In 1950, Swank observed in his patients that increased fat content was associated with a higher incidence of MS, thereby implying a potential correlation between increased fat consumption and the prevalence of MS (Swank 1950). These findings were recently supported by a pediatric MS study revealing a significant positive correlation between fat intake and disease relapse rate (Saeedeh et al. 2018).

Notably, polyunsaturated fatty acids (PUFA) and monounsaturated fatty acids (MUFA) are preferred over saturated fats in the MD. MUFA, present in key components of the MD such as olive oil, confer protective effects against oxidative damage in MS-induced rat models (Conde et al. 2020). PUFAs, abundantly present in fish and nuts, play an integral part of the MD. Plant-derived PUFAs like alpha-linolenic acid and docosahexaenoic acid exhibit anti-inflammatory properties through their transformation into prostaglandin E1 (PGE_1) and E2 (PGE_2). This transformative process subsequently leads to the downregulation of cytokine production and the migration of leukocytes (von Geldern and Mowry 2012).

Overall, epidemiological studies have consistently indicated a lower risk of MS in nations where PUFA consumption is prevalent, as opposed to those with a high intake of saturated fatty acids (Berer et al. 2018). Torkildsen et al. utilized the cuprizone model, wherein young mice are exposed to the copper chelator cuprizone through their diet, resulting in oligodendrocyte death and reversible demyelination. This model is widely used to induce toxic demyelination, particularly because spontaneous remyelination can occur as early as 4 days after cuprizone discontinuation (Torkildsen et al. 2008). Experimental findings involving cuprizone-treated mice subjected to a salmon-based diet revealed reduced demyelination, enhanced remyelination, and diminished lesion size. However, the positive impact is not universal among all types of fish, as the above findings were not reproducible in mice subjected to a cod-based diet (Torkildsen et al. 2009). Further research is therefore needed to investigate the impacts of distinct fish diets on conditions such as MS.

13.4 Human Trials

Evans et al. conducted a review of several studies to assess the effectiveness and feasibility of adhering to the MD in the context of the development and progression of MS. Their findings indicate that the various studies that explored diets in individuals with MS and animal models have a number of limitations due to the limited sample size, lack of blinding, and short durations of most human trials. Moreover, a cross-sectional retrospective study showed no correlation between adherence to the MD and the risk of developing MS (Rotstein et al. 2018). Generalizability is thus not possible and the available evidence is insufficient to make solid recommendations for a specific diet for individuals with MS (Evans et al. 2019).

One case-control study showed that increased consumption of fruits and vegetables was significantly associated with a reduced risk of developing MS (Fatemeh et al. 2016). The odds ratios were 0.28 (95% CI: 0.12–0.63, p-value: 0.002) for fruits and 0.23 (95% CI: 0.10–0.53, p-value: 0.001) for vegetables. In another cross-sectional study involving a cohort of Southern Italian patients with MS, a weak association was found between adherence to the diet and a delayed onset of MS (Esposito et al. 2021). While MD did not appear to influence the relapse rate of MS, the results suggest a favorable effect on the long-term aspects of MS, including the course of the disease and disability (Esposito et al. 2021). Considering that chronic systemic inflammation plays a vital role in autoimmunity and secondary neurodegeneration in MS, the authors hypothesized that the MD's favorable effect may stem from its potential to modulate gut microbiota and mitigate low-grade chronic systemic inflammation. This modulation, in turn, is believed to result in the downregulation of three key aspects observed in MS: sustained autoimmunity, neurodegeneration, and long-term disability (Esposito et al. 2021). In a pilot randomized controlled trial, thirty-six female participants were randomly assigned to either control or MD groups for 6 months. The study used scores to assess fatigue levels, the impact on MS symptoms, and the degree of disability caused by MS. The group following the diet demonstrated notable improvement in all three scores (Katz Sand et al. 2019). Another study investigated whether adherence to the MD could reduce inflammation factors in patients with relapsing-remitting MS (RRMS), when compared to a traditional Iranian diet. Among the 147 patients who completed the study, those adhering to the MD exhibited reduced inflammation and fatigue levels compared to those following the traditional Iranian Diet, but no significant improvement was found in disability scores (Bohlouli et al. 2022).

Ertaş et al. conducted a cross-sectional study to investigate the impact of adhering to an MD on clinical symptoms and fatigue in patients with MS. The study included 102 MS patients and used the Mediterranean diet assessment tool (MEDAS) to assess compliance. Additionally, the MS-related symptom checklist (MS-RS) and fatigue severity scale (FSS) were used to assess symptom severity in patients. While no significant correlation was found between MS-RS scores and adherence to the MD, there was compelling evidence suggesting that adherence to the MD leads to reduced fatigue (Ertaş Öztürk et al. 2023). Furthermore, various research findings emphasize the robust protective effects of the Mediterranean-style diet in mitigating MS-related fatigue compared to a conventional healthy diet. This observation remains valid even when accounting for variables such as age, BMI changes, adherence to the MD, and initial fatigue levels (Razeghi-Jahromi et al. 2020).

13.5 Therapeutic Potential of the Mediterranean Diet in Managing Multiple Sclerosis

Dietary adjustments can potentially alleviate symptoms associated with MS and reduce the severity of fatigue. The MD, touted for its anti-inflammatory components, may offer benefits in this regard; however, existing evidence in this area has been limited. Current research is exploring the correlation between adhering to a Mediterranean-style diet and the severity of MS-related symptoms and fatigue. Specific dietary alterations, such as reducing the consumption of red meat, saturated fats, and sweets while increasing the intake of fish, hold promise for mitigating MS symptoms and fatigue severity (Hassan et al. 2021). Nevertheless, these findings need to be validated through interventional studies. The MD has demonstrated positive effects on inflammation markers and appears to significantly alleviate fatigue (Hassan et al. 2021).

Further trials are needed, especially given the limited studies on this significant symptom. In addition, individuals with MS face an increased risk of developing obesity and cardiovascular diseases. It has been noted that the presence of vascular comorbidities has an adverse impact on the disability level of these patients, underscoring the importance of adopting the MD as a highly recommended measure (Hassan et al. 2021).

In terms of feasibility, the MD diet proves less restrictive compared to some alternatives and is associated with high levels of compliance (Katz Sand et al. 2019). Additionally, there are no safety concerns regarding diet intervention, as the MD is highly applicable and accessible. Demonstrating health benefits with no known unfavorable side effects, the MD offers a compelling case (Bohlouli et al. 2022). The MS community's heightened interest in the potential role of diet underscores the need for further exploration. Current data on human trials suggest that long-term MS patients can find alleviation in their symptoms by adhering to an MD course, but more robust evidence is yet to be generated (Esposito et al. 2021).

13.6 Conclusion

This chapter underscores the promising role of the MD in influencing the development and progression of MS. Emphasizing increased fruit and vegetable consumption, limited saturated fats, and avoidance of processed foods aligns with evidence suggesting a potential link between these dietary patterns and a less severe clinical course of MS. Polyunsaturated and monounsaturated fats, prevalent in the MD, as opposed to saturated fats, improve MS prognosis. While human trials present challenges, the MD's safety, accessibility, and high compliance make it a feasible dietary intervention. The limited evidence emphasizes the urgent need for comprehensive, large-scale clinical trials to elucidate the MD's therapeutic benefits in managing MS symptoms and long-term disability outcomes, addressing a critical gap in current research.

13.7 Summary

This chapter delves into the potential therapeutic benefits of the MD in managing MS, a chronic neurodegenerative disease. Emphasizing the intake of fruits, vegetables, and low saturated fats, the MD shows promise in alleviating MS symptoms through its antioxidant and anti-inflammatory effects. Increased consumption of fruits and vegetables has been linked to less severe forms of MS, attributed to their high antioxidant content. Conversely, limited saturated fats correlate with better prognosis, as they are involved in activating pro-inflammatory toll-like receptors. The chapter also highlights the protective effects of polyunsaturated and monounsaturated fats found in the MD. Despite the promising theoretical and preclinical findings, human trials have been limited and inconclusive, underscoring the need for comprehensive, large-scale studies to establish the MD's role in managing MS symptoms and long-term disability outcomes. The chapter emphasizes the MD's safety, accessibility, and high compliance, making it a viable dietary intervention for MS patients.

References

Bach-Faig A, Battino M, Belahsen R, Berry EM, Dernini S, Lairon D, Trichopoulou A (2011) Mediterranean diet pyramid today. Science and cultural updates. Public Health Nutr 14(12A):2274–2284. https://doi.org/10.1017/S1368980011002515

Berer K, Martínez I, Walker A, Kunkel B, Schmitt-Kopplin P, Walter J, Krishnamoorthy G (2018) Dietary non-fermentable fiber prevents autoimmune neurological disease by changing gut metabolic and immune status. Sci Rep 8(1):10431. https://doi.org/10.1038/s41598-018-28839-3

Bisaglia M (2022) Mediterranean diet and Parkinson's disease. Int J Mol Sci 24(1):42. https://doi.org/10.3390/ijms24010042

Bohlouli J, Borzoo-Isfahani M, Clark CCT, Hojjati Kermani MA, Moravejolahkami AR, Namjoo I, Poorbaferani F (2022) Modified Mediterranean diet v. traditional Iranian diet: efficacy of dietary interventions on dietary inflammatory index score, fatigue severity and disability in multiple sclerosis patients. Br J Nutr 128(7):1274–1284. https://doi.org/10.1017/S000711452100307X

Conde C, Escribano BM, Luque E, Aguilar-Luque M, Feijóo M, Ochoa JJ, Túnez I (2020) The protective effect of extra-virgin olive oil in the experimental model of multiple sclerosis in the rat. Nutr Neurosci 23(1):37–48. https://doi.org/10.1080/1028415X.2018.1469281

Davis C, Bryan J, Hodgson J, Murphy K (2015) Definition of the Mediterranean diet; a literature review. Nutrients 7(11):9139–9153. https://doi.org/10.3390/nu7115459

Ertaş Öztürk Y, Helvaci EM, Sökülmez Kaya P, Terzi M (2023) Is Mediterranean diet associated with multiple sclerosis related symptoms and fatigue severity? Nutr Neurosci 26(3):228–234. https://doi.org/10.1080/1028415X.2022.2034241

Esposito S, Sparaco M, Maniscalco GT, Signoriello E, Lanzillo R, Russo C, Bonavita S (2021) Lifestyle and Mediterranean diet adherence in a cohort of Southern Italian patients with Multiple Sclerosis. Mult Scler Relat Disord 47:102636. https://doi.org/10.1016/j.msard.2020.102636

Evans E, Levasseur V, Cross AH, Piccio L (2019) An overview of the current state of evidence for the role of specific diets in multiple sclerosis. Mult Scler Relat Disord 36:101393. https://doi.org/10.1016/j.msard.2019.101393

Fatemeh S, Mahsa J, Maryam B, Mostafa M, Bahram R (2016) Mediterranean diet adherence and risk of multiple sclerosis: a case-control study. Asia Pac J Clin Nutr 25(2):377–384. https://doi.org/10.6133/apjcn.2016.25.2.12

Felicetti F, Tommasin S, Petracca M, De Giglio L, Gurreri F, Ianniello A, Ruggieri S (2022) Eating hubs in multiple sclerosis: exploring the relationship between Mediterranean diet and disability status in Italy. Front Nutr 9:882426. https://doi.org/10.3389/fnut.2022.882426

Fitzgerald KC, Tyry T, Salter A, Cofield SS, Cutter G, Fox R, Marrie RA (2018) Diet quality is associated with disability and symptom severity in multiple sclerosis. Neurology 90(1):e1–e11. https://doi.org/10.1212/WNL.0000000000004768

Fotros D, Noormohammadi M, Razeghi Jahromi S, Abdolkarimi M (2023) Fruits and vegetables intake may be associated with a reduced odds of multiple sclerosis: a systematic review and dose–response meta-analysis of observational studies. Nutr Neurosci. https://doi.org/10.1080/1028415X.2023.2268390

Hadgkiss EJ, Jelinek GA, Weiland TJ, Pereira NG, Marck CH, van der Meer DM (2015) The association of diet with quality of life, disability, and relapse rate in an international sample of people with multiple sclerosis. Nutr Neurosci 18(3):125–136. https://doi.org/10.1179/1476830514Y.0000000117

Hassan A, Merghany N, Ouchkat F, Regragui W, Kedah H, Hamdy SM, Kishk NA (2021) Impact of Ramadan fasting on disease activity in patients with multiple sclerosis: a multicenter study. Nutr Neurosci. https://doi.org/10.1080/1028415X.2021.2006955

Hosseini B, Berthon BS, Saedisomeolia A, Starkey MR, Collison A, Wark PAB, Wood LG (2018) Effects of fruit and vegetable consumption on inflammatory biomarkers and immune cell populations: a systematic literature review and meta-analysis. Am J Clin Nutr 108(1):136–155. https://doi.org/10.1093/ajcn/nqy082

Huang S, Rutkowsky JM, Snodgrass RG, Ono-Moore KD, Schneider DA, Newman JW, Hwang DH (2012) Saturated fatty acids activate TLR-mediated pro-inflammatory signaling pathways. J Lipid Res 53(9):2002–2013. https://doi.org/10.1194/jlr.D029546

Katz Sand I (2018) The role of diet in multiple sclerosis: mechanistic connections and current evidence. Curr Nutr Rep 7(3):150–160. https://doi.org/10.1007/s13668-018-0236-z

Katz Sand I, Benn EKT, Fabian M, Fitzgerald KC, Digga E, Deshpande R, Arab L (2019) Randomized-controlled trial of a modified Mediterranean dietary program for multiple sclerosis: a pilot study. Mult Scler Relat Disord 36:101403. https://doi.org/10.1016/j.msard.2019.101403

Rahaman MM, Hossain R, Herrera J, Islam M, Olubunmi A, Adeyemi O, Sharifi-Rad J (2023) Natural antioxidants from some fruits, seeds, foods, natural products, and associated health benefits: an update. Food Sci Nutr. https://doi.org/10.1002/fsn3.3217

Razeghi-Jahromi S, Doosti R, Ghorbani Z, Saeedi R, Abolhasani M, Akbari N, Togha M (2020) A randomized controlled trial investigating the effects of a mediterranean-like diet in patients with multiple sclerosis-associated cognitive impairments and fatigue. Curr J Neurol 19(3):112

Rothhammer V, Mascanfroni ID, Bunse L, Takenaka MC, Kenison JE, Mayo L, Quintana FJ (2016) Type I interferons and microbial metabolites of tryptophan modulate astrocyte activity and central nervous system inflammation via the aryl hydrocarbon receptor. Nat Med 22(6):586–597. https://doi.org/10.1038/nm.4106

Rotstein DL, Cortese M, Fung TT, Chitnis T, Ascherio A, Munger KL (2018) Diet quality and risk of multiple sclerosis in two cohorts of US women. Mult Scler J 25(13):1773–1780. https://doi.org/10.1177/1352458518807061

Saeedeh A, Teri S, Jennifer G, Amy W, Anita B, Bianca Weinstock G, Emmanuelle W (2018) Contribution of dietary intake to relapse rate in early paediatric multiple sclerosis. J Neurol 89(1):28. https://doi.org/10.1136/jnnp-2017-315936

Stockinger B, Meglio PD, Gialitakis M, Duarte JH (2014) The aryl hydrocarbon receptor: multitasking in the immune system. Annu Rev Immunol 32(1):403–432. https://doi.org/10.1146/annurev-immunol-032713-120245

Swank RL (1950) Multiple sclerosis: a correlation of its incidence with dietary fat. Am J Med Sci 220:421–430

Torkildsen Ø, Brunborg LA, Myhr K-M, Bø L (2008) The cuprizone model for demyelination. Acta Neurol Scand 117(s188):72–76. https://doi.org/10.1111/j.1600-0404.2008.01036.x

Torkildsen Ø, Brunborg LA, Thorsen F, Mørk SJ, Stangel M, Myhr K-M, Bø L (2009) Effects of dietary intervention on MRI activity, de- and remyelination in the cuprizone model for demyelination. Exp Neurol 215(1):160–166. https://doi.org/10.1016/j.expneurol.2008.09.026

von Geldern G, Mowry EM (2012) The influence of nutritional factors on the prognosis of multiple sclerosis. Nat Rev Neurol 8(12):678–689. https://doi.org/10.1038/nrneurol.2012.194

Chapter 14
Ketogenic Diet: Implications on Multiple Sclerosis

Rawdah Elbahrawi, Azhar Abdukadir, Nadia Rabeh, Sara Aljoudi, Zakia Dimassi, and Hamdan Hamdan

Abstract The Ketogenic Diet (KD), renowned for treating epilepsy, is increasingly investigated for its application in Multiple Sclerosis (MS). This chapter succinctly explores recent research on the KD, focusing on its mechanisms of action in the context of MS. Emphasizing the need for further elucidation, the discussion addresses challenges in implementing the KD for MS management and the importance of ongoing research. The chapter advocates for randomized controlled trials to solidify the KD's clinical application in MS treatment. Recognizing its potential to reshape MS management, the overview concludes by highlighting the evolving landscape of KD research and its transformative implications for personalized neurological healthcare.

Keywords Multiple sclerosis (MS) · Ketogenic diet · Omega-3 · Eicosapentaenoic acid (EPA) · Docosahexaenoic acid (DHA)

Abbreviations

AcAc	Acetoacetic acid
ALA	α-linolenic acid
BBB	Blood-brain barrier

R. Elbahrawi · A. Abdukadir · N. Rabeh · S. Aljoudi · H. Hamdan (✉)
Department of Biological Sciences, College of Medicine and Health Sciences, Khalifa University, Abu Dhabi, United Arab Emirates
e-mail: hamdan.hamdan@ku.ac.ae

Z. Dimassi
Department of Medical Sciences, College of Medicine and Health Sciences, Khalifa University, Abu Dhabi, United Arab Emirates

© The Author(s), under exclusive license to Springer Nature Singapore Pte Ltd. 2024
H. Hamdan (ed.), *Exploring the Effects of Diet on the Development and Prognosis of Multiple Sclerosis (MS)*, Nutritional Neurosciences, https://doi.org/10.1007/978-981-97-4673-6_14

BHB	Beta-hydroxybutyrate
BMI	Body mass index
DHA	Docosahexaenoic acid
EAE	Experimental autoimmune encephalomyelitis
EDSS	Expanded disability status scale
EPA	Eicosapentaenoic acid
GBA	Gut-brain axis
GERD	Gastroesophageal reflux disease
IL	Interleukin
LCT	Long-chain triglyceride
LGID	Low glycemic index diet
LT	Leukotrienes
MAD	Modified Atkins diet
MS	Multiple sclerosis
NAMS	Nutritional approaches in multiple sclerosis
NLRP3	Nucleotide-binding domain, leucine-rich–containing family, pyrin domain–containing-3
PG	Prostaglandins
POLY	Polyunsaturated fat-enriched
PPARs	Peroxisome proliferator-activated receptors
PUFAs	Polyunsaturated fatty acids
ROS	Reactive oxygen species
RRMS	Relapsing-remitting MS
TX	Thromboxane

Learning Objectives
- Identify the various components and subtypes of the Ketogenic Diet (KD) and their general dietary principles.
- Explain the potential mechanisms by which the KD may influence the pathogenesis and progression of MS, focusing on anti-inflammatory and neuroprotective effects.
- Assess the current state of clinical research on the efficacy and safety of the KD in managing MS, including key findings and limitations.
- Discuss the practical challenges and potential risks associated with implementing the KD in MS patients, including issues related to adherence, side effects, and long-term sustainability.
- Propose future research directions to validate the KD's therapeutic benefits in MS, including the need for randomized controlled trials, longitudinal studies, and biomarker exploration.

14.1 Introduction

The role of diet in the treatment and prevention of disease stands as a focal point in the realm of health research (Zhu et al. 2022). An intricate interaction between dietary choices and physiological outcomes is akin to a scientific investigation, where each nutrient plays a role in orchestrating metabolic processes. Various diets exist, such as the low-carbohydrate ketogenic diet, plant-based diets, paleo diets, Mediterranean diets, and intermittent fasting, which can potentially achieve particular physiological effects (Zhu et al. 2022). One diet with a long-standing comprehensive history of its clinical use is the ketogenic diet. Evidence of its use for the treatment of epilepsy, obesity, and malignancies has demonstrated its therapeutic potential (Zhu et al. 2022). As a potential modifiable factor for the risk and progression of disease, the efficacy of the ketogenic diet in the management of multiple sclerosis (MS) has recently become the center stage of research.

MS is a neurological disease characterized by inflammation, demyelination, gliosis, and neuronal loss that comprise macro- and microscopic alterations (Di Majo et al. 2022). At the macroscopic level, focal inflammation results in plaques and injury to the blood-brain barrier (BBB), while the microscopic level manifests through the impairment of axons, neurons, and synapses (Di Majo et al. 2022). Due to its multifactorial nature, the precise etiology of MS is not entirely understood. A proposed mechanism of MS development is gut dysbiosis, a disruption in the gut microbiome composition (Miyake et al. 2015). The changes seen in the gut can have trickling effects on the nervous system. This relationship is coined the gut-brain axis (GBA) and is also influenced by various organ systems, such as the endocrine and immune systems. Thus, the interplay between diet, physiology, and the gut microbiome is thought to be a determining factor triggering inflammation in MS (Di Majo et al. 2022; Esposito et al. 2018). This chapter aims to shed light on the components of the ketogenic diet, its role in neuroinflammation, evidence of its use for the management of MS, its potential risks, and future areas of research.

14.2 Components of a Ketogenic Diet

The ketogenic diet is a low-carbohydrate diet that essentially results in ketosis, elevation of fatty acids, and modulation of glycemia. It is generally based on increased protein and fat intake, which forces the body to produce and release large amounts of ketone bodies (Di Majo et al. 2022). The term ketogenic diet is an umbrella term for its various subtypes: long-chain triglyceride and medium-chain triglyceride, ketogenic diets, modified Atkins diet (MAD), and low glycemic index diet (LGID). These subtypes are generally effective, but whether they act through the same or different mechanisms of action is yet to be determined (Zhu et al. 2022).

The long-chain triglyceride (LCT) ketogenic diet is the oldest form of the ketogenic diet. The regimen consists of a 4:1 ratio of fats to carbohydrates and protein, with approximately 90% of calories derived from fats. In patients or infants with a higher demand for protein intake to sustain growth, a reduced ratio of 3:1 can still be effective in maintaining a state of ketosis (Zhu et al. 2022). Despite evidence of this diet in treating certain diseases, it is often difficult to adhere to. Challenges arise from difficulties in preparing LCT ketogenic-friendly meals, which are often unpalatable (Zhu et al. 2022). A more reasonable alternative is the medium-chain triglyceride (MCT) ketogenic diet.

The second oldest regimen is the MCT ketogenic diet, which uses MCT chains of 6–12 carbons in length (Zhu et al. 2022). Not only are these fats more palatable, but they were found to be more highly ketogenic than the fats consumed in the LCT ketogenic diet. Due to the elevated ketogenic potential of MCTs, individuals can consume a lower amount of fats and increase their carbohydrate and protein intake (Zhu et al. 2022).

The MAD, the next ketogenic diet published in 2003, became increasingly attractive as it is less restrictive than other regimens (Zhu et al. 2022). Geared toward individuals with behavioral problems and children, the MAD derives 65% of its calories from fats. Generally, the MAD is accepted and better tolerated than the classic LCT ketogenic diet (Zhu et al. 2022).

Lastly, the LGID introduced in 2005 builds upon previous ketogenic diets, but it is free of carbohydrate intake restrictions. The only restriction of the LGID is that consumed foods should have a low glycemic index, preferably below 50% (Zhu et al. 2022). The glycemic index is an incremental measurement of how much an item of food raises an individual's blood glucose levels. It provides a relative ranking of carbohydrates based on their immediate effect on blood sugar when consumed. The scale is typically measured from 0 to 100, with higher values indicating a more rapid and significant increase in blood glucose levels (Trumbo 2021).

The type of fats consumed as part of the ketogenic diet plays a central role in its effect on the body. One important group is the polyunsaturated fatty acids (PUFAs) comprised of omega-3 and omega-6 (Di Majo et al. 2022). Omega-6 is found in various animal products and vegetable oils, producing pro-inflammatory effects (Di Majo et al. 2022). On the other hand, omega-3 can indirectly regulate transcription factors involved in the expression of inflammatory genes, influence the composition of gut microbiota, and enhance the production of anti-inflammatory compounds (Di Majo et al. 2022; Taha et al. 2010). Three types of essential omega-3 fatty acids that can only be introduced through the diet are recognized in nutrition; these are α-linolenic acid (ALA), eicosapentaenoic acid (EPA), and docosahexaenoic acid (DHA). ALA is found in vegetable oils, while EPA and DHA are found in fish (Di Majo et al. 2022). DHA is a crucial component of cell membranes and is present in high concentrations in the retina and the brain (Di Majo et al. 2022). EPA is a critical factor in the synthesis of eicosanoids and competes with arachidonic acid to produce prostaglandins (PG), thromboxane (TX), and leukotrienes (LT). A higher concentration of EPA drives the synthesis of eicosanoids with less inflammatory activity (Di Majo et al. 2022). Therefore, it is possible that a ketogenic diet focused on increasing fat and protein intake can play a role in mitigating the inflammatory process in diseases such as MS.

14.3 Neuroinflammation and Ketogenic Diet

The efficacy of the ketogenic diet for MS management is said to be based on the reduction in inflammatory markers (Bock et al. 2018; Lee et al. 2021). The ketogenic diet modulates inflammation by producing ketones, namely beta-hydroxybutyrate (BHB) and acetoacetic acid (AcAc), which act by inhibiting mitochondrial permeability and enhancing NADH oxidation. More importantly, these ketones exert a neuroprotective effect by reducing oxidative stress and reactive oxygen species (ROS) (Di Majo et al. 2022). Other inflammatory pathways that ketone bodies target include the peroxisome proliferator-activated receptors (PPARs), the activation of which can reduce inflammation, as seen in mice models. Of the PPAR expressed in the body, PPAR-γ is expressed in the brain, hence its importance in the modulation of neuroinflammation (Koh et al. 2020). A promising finding in the application of the ketogenic diet is its ability to promote axon remyelination (Ortí et al. 2023; Storoni and Plant 2015). Reductions in serum neurofilament levels, which usually cause axonal damage in patients with MS, have also been reported (Bock et al. 2021; Ortí et al. 2023).

The nucleotide-binding domain, leucine-rich–containing family, pyrin domain–containing-3 (NLRP3) inflammasome has recently been considered a critical factor in the development of neuroinflammation. Recent studies show that NLRP3 acts as a bridge between the innate and adaptive immune response in the initial stages of MS by promoting the migration of macrophages, dendritic cells, and myelin-specific autoreactive CD4+ T cells (Gharagozloo et al. 2018; Yang et al. 2013). The antiinflammatory mediator, BHB, has exhibited an inhibitory effect on NLRP3, downregulating the inflammasome-mediated activation of pro-inflammatory interleukin (IL)-1B and IL-18 cytokines from macrophages (Youm et al. 2015). Ultimately, these new findings offer a new potential management and therapeutic target for MS.

14.4 Ketogenic Diet and MS

Owing to its inhibitory effects on the production of inflammatory markers, the ketogenic diet has been shown to reduce disease progression in studies on mice with experimental autoimmune encephalomyelitis (EAE), an experimentally induced form of MS (Schwarz and Leweling 2005; Yao et al. 2021). The effects of the ketogenic diet are substantial, and remyelination has been observed in EAE mice models (Liu et al. 2020). Despite the accumulation of promising evidence, the number of clinical trials exploring the therapeutic effect of the ketogenic diet on MS progression is limited (Bahr et al. 2020; Nicholas Brenton et al. 2022; Brenton et al. 2022).

The ketogenic diet's efficacy in treating MS patients was assessed in a pilot study on a cohort of patients with relapsing-remitting MS. This open-label, single-arm trial recruited 20 patients who were instructed to adhere to the MAD for 6 months

(Brenton et al. 2019). The trial found a 95% adherence to the diet at 3 months, but this dropped to 75% at 6 months. Despite a reduction in adherence, patients generally reported improved body mass index (BMI), depression, and fatigue scores. More importantly, disease progression or serious adverse effects were not observed (Brenton et al. 2019). Arguably, the results should be interpreted with caution as the study did not have a control group or a large sample size. Nonetheless, the positive outcome of the pilot study motivated clinical trials to pursue this diet as a possible intervention by the same researchers who pursued a phase II trial assessing the efficacy and tolerability of the ketogenic diet in relapsing-remitting MS (RRMS) patients (Nicholas Brenton et al. 2022). Sixty-five participants were enrolled and instructed to follow a MAD for 6 months. The trial observed the same benefits to the diet as in their pilot study. Eighty-three percent (83%) of participants adhered to the regimen by the end of the trial. Along with reductions in fat, fatigue, and depression, other measures of disease severity, such as the Expanded Disability Status Scale (EDSS) scores, were improved by the end of the trial (Nicholas Brenton et al. 2022). This trial was the first of its kind to demonstrate the efficacy and tolerability of the ketogenic diet for MS patients.

Once the safety of the diet was ensured, additional clinical trials and their results were rolled out. A randomized, parallel-group clinical trial named NAMS (Nutritional Approaches in Multiple Sclerosis) investigated the effect of the ketogenic and fasting diet in patients with MS (Bahr et al. 2020). The trial's results are still pending, but what distinguishes this trial from the pioneering ones is its investigation of changes to the gut microbiome and indicators of autophagy, inflammation, and oxidative stress (Bahr et al. 2020). In another clinical trial evaluating the sustainability of the ketogenic diet, 65 participants were enrolled and instructed to follow the diet for 6 months (Wetmore et al. 2023). Three months after completing the intervention, participants were asked to conduct a post-trial follow-up to evaluate the participants' likelihood of continuing the regimen. Eighty-one percent of the participants returned for the 3-month post-trial follow-up. Of these participants, 21% continued to strictly adhere to the ketogenic diet, while 37% continued following the diet in a less restrictive manner (Wetmore et al. 2023). The study participants reported a reduction in BMI and fatigue, and these results were sustained when the ketogenic diet was continued after the trial. An interesting finding is that participants who experienced a greater reduction in BMI and fatigue were more likely to follow a strict form of the ketogenic diet after the completion of the trial (Wetmore et al. 2023). Essentially, adherence improved when patients noted significant improvements in their habitus. This is a noteworthy finding, as it is thought that symptom-free patients often show reduced levels of adherence to their treatment (Jimmy and Jose 2011).

Understanding the effects of the ketogenic diet on MS activity was previously attempted through clinical trials that lacked the measurement of particular biomarkers as a proxy for neuroaxonal injury. Oh et al. conducted a clinical trial aimed at measuring neurofilament levels, a biomarker that acts as a proxy measure of neuroaxonal damage, in a cohort of 39 patients with relapsing MS (Oh et al. 2023). The participants were divided into a treatment group and a control group, where the treatment group was instructed to follow a ketogenic diet for 6 months. The authors did not provide sufficient details on the subtype of the ketogenic diet the participants adhered to. The study found that the diet did not induce further neuroaxonal injury by the relatively stable levels of neurofilament (Oh et al. 2023). A noteworthy observation was that patients exhibiting a greater degree of ketosis, especially higher levels of BHB, had significantly improved neurofilament levels (Oh et al. 2023). Additional research is needed to understand the significance of the type of ketones produced and their relationship with neuroaxonal injuries.

14.5 Dietary Recommendations

Incorporating dietary changes in MS management plans has been of widespread interest. Effects of diet on MS can possibly be the consequence of the direct action of metabolites produced by food, the effect of the metabolites that are synthesized by the gut microflora, or possibly an alteration in gut microbial composition due to dietary changes (Di Majo et al. 2022). An example nutritional plan proposed by Di Majo et al. (2022) includes two steps to approaching a ketogenic diet protocol: an adaptation phase and a maintenance phase. The adaptation phase is further divided into two periods in which, in the first 4 weeks, the patients should be instructed to limit their intake of carbohydrates to 20 g/day; this is done to establish a state of ketosis. During the second period, the patient will increase their carbohydrate intake by 5 g each week until a maximum of 40 g/day. The maintenance phase aims to maintain a constant state of nutritional ketosis, with ketone body levels between 0.5 and 3 mM.

Further, they refer to a clinical trial by Fuehrlein et al. (2004) that followed a similar approach to their proposed nutritional plan (Fuehrlein et al. 2004). The trial recruited twenty healthy adults and randomized them into two different weight-maintaining ketogenic diets for 5 days. The diet breakdown was 70% fat, 15% carbohydrate, and 15% protein. Additionally, the fat contents were 60 or 15 % saturated, 15 or 60% polyunsaturated, and 25% monounsaturated for the saturated fatty acid-enriched ketogenic diet and polyunsaturated fat-enriched (POLY) ketogenic diet. They measured changes in serum BHB, insulin sensitivity, and lipid profiles. They found that a short-term POLY ketogenic diet induces a greater level of ketosis ($p = 0.0004$) and improves insulin sensitivity ($p = 0.02$) without negatively impacting the total and low-density lipoprotein cholesterol. Although their findings indicate that a POLY ketogenic diet can significantly alter the ketosis state, more research is required as their sample size was too small. The duration of the study was not long. Additional research is required to see any association or changes in individuals with MS. Bahr et al. (2020) are conducting a randomized, controlled, parallel-group study that enlisted 111 patients with relapsing-remitting type MS who received immunomodulatory or did not receive disease-modifying therapy. The participants were then randomized to one of three 18-month dietary interventions. The dietary interventions of interest include a ketogenic diet with a restricted carbohydrate intake of 20–40 g/day, a fasting diet with a 7-day fast every 6 months and a 14-h daily intermittent fast in between, and a fat-modified standard diet as recommended by the German Nutrition Society (Bahr et al. 2020). Unfortunately, the results of the clinical trial have not been published. Given the limited evidence regarding the regimen's therapeutic potential, additional research is needed to overcome this knowledge gap.

14.6 Potential Risks and Concerns

During the initial weeks of the ketogenic diet, individuals commonly experience short-term adverse effects, including irritability, headache, metabolic acidosis, fatigue, dehydration, diarrhea, nausea, hypoglycemia, and reluctance to eat (Zhu et al. 2022). These predictable and preventable responses are typically a result of diet-induced metabolic shifts. In the long term, the ketogenic diet may lead to adverse effects such as heightened cardiovascular risks, indicated by unfavorable cholesterol profiles and nephrolithiasis (Zhu et al. 2022). This is likely associated with metabolic effects such as hypercalciuria, decreased bone mineral density, growth retardation, anemia, hypocitraturia, and neuropathy (Crosby et al. 2021). Throughout ketogenic diet therapy, individuals often face gastrointestinal disturbances, including constipation, gastroesophageal reflux disease (GERD), vomiting, and abdominal pain (Zarnowska 2020; Zhu et al. 2022).

Given that the ketogenic diet is a multidisciplinary therapy, successful implementation requires the active participation of experienced caregivers. An effective collaboration between dietitians, nurses, and doctors is crucial to implement holistic patient care (Howrie et al. 1998). Despite the cooperative efforts to deliver patient-centered care, challenges related to treatment adherence persist. Some factors that may lead to non-adherence include patient food preferences and the unpleasant taste of most ketogenic meals. The restrictive nature of the diet can also be a factor that prompts patients to abandon or avoid the diet. These factors may diminish patient motivation, leading to discontinuation and subsequent treatment failure (Howrie et al. 1998; Zhu et al. 2022).

14.7 Future Directions

Our current understanding of the ketogenic diet's therapeutic potential stems from extensive preclinical research. Preclinical research is indispensable and has informed our understanding of the effects of the ketogenic diet on the body. However, a shift from preclinical to clinical research is needed to understand the diet's potential in patients. Previous clinical trials have established the diet as tolerable and safe. Clinical trials have also highlighted patient adherence to the diet in cases of significant improvement in disease activity. An extension of these studies and findings is to assess the efficacy of utilizing different regimens simultaneously. For example, many patients find the classic ketogenic diets, such as the LCT and MCT diets, unpalatable or restrictive. Therefore, it would be worth investigating if a combination of LCT or MCT can be made with more tolerable diets such as the MAD or LGID. Additionally, clinical trials investigating the effect of diet and pharmacological therapies on MS are yet to be executed. Such trials can determine if diet enhances the effects of pharmacological agents, and vice versa, for MS treatment.

The clinical trials discussed often restrict the duration of the diet to 6 months. Longitudinal trials capturing the full extent of the diet's efficacy are lacking. Essential findings from longitudinal studies can enhance our understanding of the diet's impact throughout the disease course. In addition to monitoring the effect of dietary changes on MS activity, it is essential to explore the effect of the diet in other forms of MS, as previous clinical trial cohorts were commonly of the RRMS subtype.

The clinical trial investigating the effect of the ketogenic diet on neurofilament levels in MS patients was the first of its kind to measure a biomarker for neurodegeneration. Clinical trials remain needed to explore and understand the mechanism by which the ketogenic diet influences aspects such as oxidative stress, autophagy, and neuroinflammation. The lack of such trials can be attributed to the lack of standard protocols or biomarkers to monitor disease progress. In recent years, promising biomarkers for use in clinical practice have emerged (Mathur et al. 2021). However, given the unpredictable course of MS, it is increasingly difficult to find a single, unique diagnostic biomarker. Currently, research efforts should focus on discovering a panel of biomarkers for MS diagnosis and prognosis.

14.8 Conclusion

In conclusion, the ketogenic diet has gained attention as a potential complementary approach to managing MS symptoms and progression. While it shows promise, further research, including long-term studies, is needed to better understand its benefits and risks in the context of MS. The ketogenic diet reduces inflammation, enhances cognitive function, and promotes neuroprotection. Therefore, Individuals with MS interested in the ketogenic diet should consult with healthcare providers to develop a personalized dietary plan that aligns with their health goals and needs.

14.9 Summary

This chapter explores the potential of the KD as a therapeutic approach for managing MS. The KD, traditionally used to treat epilepsy, is gaining attention for its possible benefits in MS due to its anti-inflammatory and neuroprotective properties. The diet, which emphasizes high fat and low carbohydrate intake, induces ketosis, potentially reducing neuroinflammation and oxidative stress. While preliminary studies in animal models and small-scale human trials show promise, there are significant challenges in implementing the KD for MS management, including adherence difficulties and potential side effects. The chapter underscores the necessity for extensive randomized controlled trials and longitudinal studies to better understand the KD's efficacy, safety, and mechanisms of action in MS. By addressing these research gaps, the KD could become a valuable addition to personalized neurological healthcare for MS patients.

References

Bahr LS, Bock M, Liebscher D, Bellmann-Strobl J, Franz L, Prüß A, Schumann D, Piper SK, Kessler CS, Steckhan N, Michalsen A, Paul F, Mähler A (2020) Ketogenic diet and fasting diet as nutritional approaches in multiple sclerosis (NAMS): protocol of a randomized controlled study. Trials 21(1):3. https://doi.org/10.1186/s13063-019-3928-9

Bock M, Karber M, Kuhn H (2018) Ketogenic diets attenuate cyclooxygenase and lipoxygenase gene expression in multiple sclerosis. EBioMedicine 36:293–303. https://doi.org/10.1016/j.ebiom.2018.08.057

Bock M, Steffen F, Zipp F, Bittner S (2021) Impact of dietary intervention on serum neurofilament light chain in multiple sclerosis. Neurol Neuroimmunol Neuroinflam 9(1):e1102. https://doi.org/10.1212/NXI.0000000000001102

Brenton JN, Banwell B, Bergqvist AGC, Lehner-Gulotta D, Gampper L, Leytham E, Coleman R, Goldman MD (2019) Pilot study of a ketogenic diet in relapsing-remitting MS. Neurol Neuroimmunol Neuroinflam 6(4):e565. https://doi.org/10.1212/NXI.0000000000000565

Brenton JN, Diana L-G, Emma W, Brenda B, Bergqvist AGC, Shanshan C, Rachael C, Mark C, Myla DG (2022) Phase II study of ketogenic diets in relapsing multiple sclerosis: safety, tolerability and potential clinical benefits. J Neurol 93(6):637. https://doi.org/10.1136/jnnp-2022-329074

Crosby L, Davis B, Joshi S, Jardine M, Paul J, Neola M, Barnard ND (2021) Ketogenic diets and chronic disease: weighing the benefits against the risks. Front Nutr 8:702802

Di Majo D, Cacciabaudo F, Accardi G, Gambino G, Giglia G, Ferraro G, Candore G, Sardo P (2022) Ketogenic and modified Mediterranean diet as a tool to counteract neuroinflammation in multiple sclerosis: nutritional suggestions. Nutrients 14(12):2384

Esposito S, Bonavita S, Sparaco M, Gallo A, Tedeschi G (2018) The role of diet in multiple sclerosis: a review. Nutr Neurosci 21(6):377–390. https://doi.org/10.1080/1028415X.2017.1303016

Fuehrlein BS, Rutenberg MS, Silver JN, Warren MW, Theriaque DW, Duncan GE, Stacpoole PW, Brantly ML (2004) Differential metabolic effects of saturated versus polyunsaturated fats in ketogenic diets. J Clin Endocrinol Metabol 89(4):1641–1645. https://doi.org/10.1210/jc.2003-031796

Gharagozloo M, Gris KV, Mahvelati T, Amrani A, Lukens JR, Gris D (2018) NLR-dependent regulation of inflammation in multiple sclerosis. Front Immunol 8:2012. https://doi.org/10.3389/fimmu.2017.02012

Howrie DL, Kraisinger M, McGhee HW, Crumrine PK, Katyal N (1998) The ketogenic diet: the need for a multidisciplinary approach. Ann Pharmacother 32(3):384–385. https://doi.org/10.1345/aph.17201

Jimmy B, Jose J (2011) Patient medication adherence: measures in daily practice. Oman Med J 26(3):155–159. https://doi.org/10.5001/omj.2011.38

Koh S, Dupuis N, Auvin S (2020) Ketogenic diet and Neuroinflammation. Epilepsy Res 167:106454. https://doi.org/10.1016/j.eplepsyres.2020.106454

Lee JE, Titcomb TJ, Bisht B, Rubenstein LM, Louison R, Wahls TL (2021) A modified MCT-based ketogenic diet increases plasma β-hydroxybutyrate but has less effect on fatigue and quality of life in people with multiple sclerosis compared to a modified paleolithic diet: a waitlist-controlled, randomized pilot study. J Am Coll Nutr 40(1):13–25. https://doi.org/10.1080/07315724.2020.1734988

Liu C, Zhang N, Zhang R, Jin L, Petridis AK, Loers G, Zheng X, Wang Z, Siebert H-C (2020) Cuprizone-Induced demyelination in mouse hippocampus is alleviated by ketogenic diet. J Agric Food Chem 68(40):11215–11228. https://doi.org/10.1021/acs.jafc.0c04604

Mathur D, Mishra BK, Rout S, Lopez-Iranzo FJ, Lopez-Rodas G, Vallamkondu J, Kandimalla R, Casanova B (2021) Potential biomarkers associated with multiple sclerosis pathology. Int J Mol Sci 22(19):10323

Miyake S, Kim S, Suda W, Oshima K, Nakamura M, Matsuoka T, Chihara N, Tomita A, Sato W, Kim S-W, Morita H, Hattori M, Yamamura T (2015) Dysbiosis in the gut microbiota of patients with multiple sclerosis, with a striking depletion of species belonging to clostridia XIVa and IV clusters. PLoS ONE 10(9):e0137429. https://doi.org/10.1371/journal.pone.0137429

Oh U, Woolbright E, Lehner-Gulotta D, Coleman R, Conaway M, Goldman MD, Brenton JN (2023) Serum neurofilament light chain in relapsing multiple sclerosis patients on a ketogenic diet. Mult Scler Relat Disord 73:104670. https://doi.org/10.1016/j.msard.2023.104670

Ortí JE, Cuerda-Ballester M, Sanchis-Sanchis CE, Lajara Romance JM, Navarro-Illana E, García Pardo MP (2023) Exploring the impact of ketogenic diet on multiple sclerosis: obesity, anxiety, depression, and the glutamate system. Front Nutr 10:1227431. https://doi.org/10.3389/fnut.2023.1227431

Schwarz S, Leweling H (2005) Multiple sclerosis and nutrition. Mult Scler J 11(1):24–32. https://doi.org/10.1191/1352458505ms1119oa

Storoni M, Plant GT (2015) The therapeutic potential of the ketogenic diet in treating progressive multiple sclerosis. Mult Scler Int 2015:681289. https://doi.org/10.1155/2015/681289

Taha AY, Burnham WM, Auvin S (2010) Polyunsaturated fatty acids and epilepsy. Epilepsia 51(8):1348–1358. https://doi.org/10.1111/j.1528-1167.2010.02654.x

Trumbo PR (2021) Global evaluation of the use of glycaemic impact measurements to food or nutrient intake. Public Health Nutr 24(12):3966–3975. https://doi.org/10.1017/S1368980021000616

Wetmore E, Lehner-Gulotta D, Florenzo B, Banwell B, Bergqvist AGC, Coleman R, Conaway M, Goldman MD, Brenton JN (2023) Ketogenic diet in relapsing multiple sclerosis: patient perceptions, post-trial diet adherence outcomes. Clin Nutr 42(8):1427–1435. https://doi.org/10.1016/j.clnu.2023.06.029

Yang Y, Inatsuka C, Gad E, Disis ML, Standish LJ, Pugh N, Pasco DS, Lu H (2013) Protein-bound polysaccharide-K induces IL-1β via TLR2 and NLRP3 inflammasome activation. Innate Immun 20(8):857–866. https://doi.org/10.1177/1753425913513814

Yao A, Li Z, Lyu J, Yu L, Wei S, Xue L, Wang H, Chen G-Q (2021) On the nutritional and therapeutic effects of ketone body d-β-hydroxybutyrate. Appl Microbiol Biotechnol 105(16):6229–6243. https://doi.org/10.1007/s00253-021-11482-w

Youm Y-H, Nguyen KY, Grant RW, Goldberg EL, Bodogai M, Kim D, D'Agostino D, Planavsky N, Lupfer C, Kanneganti TD, Kang S, Horvath TL, Fahmy TM, Crawford PA, Biragyn A, Alnemri E, Dixit VD (2015) The ketone metabolite β-hydroxybutyrate blocks NLRP3 inflammasome–mediated inflammatory disease. Nat Med 21(3):263–269. https://doi.org/10.1038/nm.3804

Zarnowska IM (2020) Therapeutic use of the ketogenic diet in refractory epilepsy: what we know and what still needs to be learned. Nutrients 12(9):2616

Zhu H, Bi D, Zhang Y, Kong C, Du J, Wu X, Wei Q, Qin H (2022) Ketogenic diet for human diseases: the underlying mechanisms and potential for clinical implementations. Signal Transduct Target Ther 7(1):11. https://doi.org/10.1038/s41392-021-00831-w

Chapter 15
Epigenetics: Implication on Multiple Sclerosis

Rawdah Elbahrawi, Sara Aljoudi, Nadia Rabeh, Zakia Dimassi, Khalood Mohamed Alhosani, and Hamdan Hamdan

Abstract The pathogenesis of Multiple sclerosis is not singly identified or entirely understood; however, it is hypothesized to be a multifactorial acquired disease influenced by environmental, genetic, and epigenetic modifications. Epigenetics results in altered gene expression without altering the genetic code and does so by three primary mechanisms: DNA methylation, histone code modifications, and miRNA-regulated gene expression. DNA methylation adds methyl groups to cytosine residues, primarily at CpG dinucleotides, altering chromatin structure to silence genes or repress transcription. Meanwhile, histone code modifications, including acetylation, methylation, phosphorylation, and ubiquitination of histone tails, can limit chromatin accessibility and gene expression. Additionally, miRNAs, small noncoding RNAs, post-transcriptionally regulate gene expression by binding to target mRNAs, either inhibiting translation or promoting mRNA degradation. Dysregulation of these epigenetic mechanisms is implicated in various diseases, such as multiple sclerosis (MS), emphasizing the importance of understanding their intricate interplay in disease development and progression.

Keywords Multiple sclerosis (MS) · Epigenetic · Gene regulation · Environmental influences · Genetics

---

R. Elbahrawi · S. Aljoudi · N. Rabeh · K. M. Alhosani · H. Hamdan (✉)
Department of Biological Sciences, College of Medicine and Health Sciences, Khalifa University, Abu Dhabi, United Arab Emirates
e-mail: hamdan.hamdan@ku.ac.ae

Z. Dimassi
Department of Medical Sciences, College of Medicine and Health Sciences, Khalifa University, Abu Dhabi, United Arab Emirates

© The Author(s), under exclusive license to Springer Nature Singapore Pte Ltd. 2024
H. Hamdan (ed.), *Exploring the Effects of Diet on the Development and Prognosis of Multiple Sclerosis (MS)*, Nutritional Neurosciences, https://doi.org/10.1007/978-981-97-4673-6_15

Abbreviations

(25 (OH)D)	25-Hydroxyvitamin D
CI	Confidence interval
CIS	Clinically isolated syndrome
CNS	Central nervous system
EBV	Epstein Barr virus
EDSS	Expanded disability status scale
GRS	Genetic risk scores
MBP	Myelin basic protein
miRNA	MicroRNAs
MS	Multiple sclerosis
NF-κB	Nuclear factor kappa B
PAD2	Peptidyl arginine deaminase 2
SIR	Standardized incidence ratio
Th	T helper
Tregs	Regulatory T
UVB	Ultraviolet B radiation

Learning Objectives
- Describe the three primary mechanisms of epigenetic regulation: DNA methylation, histone code modifications, and microRNA-regulated gene expression, and their roles in altering gene expression without changing the DNA sequence.
- Explain the process of DNA methylation, its role in gene silencing, and how it contributes to chromatin structure and gene repression.
- Identify different types of histone modifications, such as acetylation and methylation, and understand how these modifications influence chromatin accessibility and gene expression.
- Discuss the function of microRNAs in posttranscriptional gene regulation, their impact on mRNA stability and translation, and their relevance in the pathogenesis of MS.
- Analyze how dysregulation of epigenetic mechanisms contributes to the development and progression of MS and explore potential therapeutic strategies targeting these epigenetic modifications.

15.1 Introduction

The study of epigenetics encompasses an appreciation of how genetic traits can result in steadily heritable phenotypes through changes in gene expression without altering the underlying DNA sequence. This process is characterized by modifications in gene expression rather than changes to the genetic code (Berger et al. 2009). Mechanisms of epigenetics regulate gene expression by controlling the accessibility of transcriptional machinery to the regulatory regions of genes (van den Elsen et al. 2014). In the context of health, multiple sclerosis (MS), an autoimmune demyelinating, neurodegenerative disease affecting the central nervous system (CNS), is a major cause of morbidity, impacting between 50–300 individuals per 100,000, with an increasing incidence and prevalence in both developed and developing countries (Adamczyk-Sowa et al. 2020; Browne et al. 2014). MS exhibits diverse phenotypic presentations, categorized into various subtypes, such as clinically isolated syndrome (CIS), relapsing-remitting, primary progressive, secondary progressive, and progressive-relapsing (Di Majo et al. 2022; Lublin 2014). CIS, accounting for the initial presentation in 80% of MS cases, involves acute clinical attacks affecting one or more CNS sites (Doshi and Chataway 2016). Typically, clinical manifestations of MS are related to its involvement in various systems, such as motor, sensory, visual, and autonomic systems (Doshi and Chataway 2016).

15.2 Etiology of Multiple Sclerosis

At the macroscopic level, focal inflammation leads to the formation of plaques, which are the pathognomonic lesions of MS, as well as injury to the blood-brain barrier. On the microscopic level, neurodegeneration impairs axons, neurons, and synapses, activating a robust autoimmune response involving macrophages, T and B cells, and cytotoxic microglia (Häusser-Kinzel and Weber 2019). The clinical presentation of MS often assumes a relapsing and remitting pattern characterized by episodes of neurological dysfunction, after which patients experience recovery but do not return to their baseline normal functions. This pattern is hypothesized to be a result of transient inflammation that is followed by remyelination, albeit not a robust one (Compston et al. 2008). Over time, the cumulative impact of disability and incomplete recovery from each relapse contributes to the progression of the disease.

Although the root origin of MS is not fully understood, it is theorized to be influenced by genetics, environmental elements, and epigenetic modifications. All of these factors collectively shape the complex manifestations and progression of MS.

15.2.1 *Genetic Influence on Multiple Sclerosis*

Research findings indicate a crucial role of genetic factors in the increased frequency of MS among relatives of affected individuals (Dyment et al. 2006; Hemminki et al. 2009). Hemminki et al. (2009) utilized the Multigeneration Register in Sweden and the Hospital Discharge Register to evaluate familial MS risk in various relationships. This assessment considered cases where a family member was diagnosed with MS or one of 33 other autoimmune diseases, including ankylosing spondylitis and Graves disease. Among the total patient population of 42,510,211,154 were diagnosed with MS. They calculated the standardized incidence ratio (SIR) for family members of patients with MS, comparing those with and without affected family members. Results revealed that the SIR for MS was 5.94 [95% confidence interval (CI) [4.88–7.16] in offspring of affected parents, 6.25 [95% CI [3.66, 10.60]] in singleton siblings, 9.09 [95% CI [1.67, 33.24] in twins, and 1.50 [95% CI [0.93, 2.30]; insignificant] in spouses. The overall risk of MS was 1.21 [95% CI [1.14–1.28]] when a parent was diagnosed with any autoimmune disease, implicating genetics in MS risk (Hemminki et al. 2009). In addition, familial clustering, or the occurrence of the disease in excess within some families that is greater than expected from the occurrence in the general population, has demonstrated that first-degree relatives of probands have a 30–50 times greater MS risk than 0.1% of the general population (Borch-Johnsen et al. 1994; Sadovnick et al. 1988). Twin studies further support partial genetic susceptibility to MS, revealing a significant twofold increase in the risk of developing MS in dizygotic twins over nontwin siblings of twins (Willer et al. 2003). Despite abundant research on genetic ties to MS, a consensus among physicians and scientists suggests that the etiology is grounded in a multifactorial pathway.

15.2.2 Environmental Influence on Multiple Sclerosis

15.2.2.1 Latitude

There are documented findings that have detected a possible connection between differing latitude habitation and the risk of developing MS. A majority of MS cases are reported in temperate regions where sunlight is less potent (Donnan et al. 2005; Ramagopalan et al. 2011). Simpson et al. explored the role of ultraviolet B radiation (UVB) in explaining the period prevalence of MS in England, which is a measure of the proportion of individuals in England with MS during a period of time. This study included all admissions to the National Health Service hospitals in England from 1998 to 2005 along with data on UVB intensity for England from the Nimbus 7 satellite from the United States National Aeronautics and Space Administration. Results suggest that the regression of MS against UVB intensity for all seasons has an r^2 of 0.61, which suggests a moderately strong positive linear relationship between MS prevalence and UVB intensity. However, this study is limited to a certain population and did not take into consideration the varying exposure to UVB between patients. Simpson Jr et al. performed a meta-regression, confirming a global association between MS prevalence and latitude ($p = 0.001$) (Simpson et al. 2011). While several studies explore latitude as a risk factor for MS, such as those by Vitkova et al. (2022), further research is needed to understand the association between latitude and MS prevalence, particularly in non-European/Western countries (Ostkamp and Schwab 2022).

15.2.2.2 Vitamin D

Decreased serum vitamin D levels in cases of MS were first evidenced by Nieves et al. in 1994. The role of Vitamin D in the pathogenesis of MS has been well-examined, with multiple studies providing strong evidence for the correlation between low vitamin D levels and the risk of developing MS (Nieves et al. 1994; Tarlinton et al. 2019). A study conducted by Brola et al. in 2014 aimed to evaluate the association of serum 25-Hydroxyvitamin D (25 (OH)D) with disability and frequency of relapses in relapsing-remitting MS. The study included 184 patients diagnosed with relapsing-remitting MS, undergoing immune-modulation drug treatments but not receiving vitamin D supplementation. To assess the impact of these interventions, the study examined the levels of disability experienced by the patients using the Expanded Disability Status Scale (EDSS). They divided the patients into two groups based on the EDSS score: those between 0.0 and 2 (mild) in one group and those between 2.5 and 4 (severe) in another. Levels of serum 25(OH)D were then compared with the occurrence of relapses and the level of disability. The study findings indicate significantly lower mean serum concentration of 25(OH)D during winter in both MS patients and controls. Specifically, serum levels were found to be significantly lower in severe cases compared to mild ones and controls ($p = 0.022$ and $p = 0.008$, respectively). Logistic regression revealed a significant association between winter serum levels of 25(OH)D and MS (odds ratio 0.925, 95% CI [0.822–0.970]), indicating that low vitamin D levels are associated with severe MS and increased relapse rates (Brola et al. 2016). Another study employed Mendelian randomization to assess the causal association between low serum vitamin D levels, increased body mass index, and Pediatric-onset MS, using genetic risk scores [GRS], which are numerical summaries representing an individual's genetic predisposition to a particular disease based on their genetic profile (Gianfrancesco et al. 2017). The study recruited 294 MS cases and 10,875 controls in the United States, along with 175 cases and 5376 controls from Sweden. Their meta-analysis revealed that higher vitamin D GRS, associated with increased serum levels of 25(OH)D, significantly decreased the odds of Pediatric-onset MS. The calculated odds ratio was 0.72, accompanied by a 95% CI of [0.55, 0.94] and a p-value of 0.02 (Gianfrancesco et al. 2017). Multiple studies have been done to assess the association of serum vitamin D levels with MS. However, it would be interesting to assess the value of screening at-risk populations, such as women and Caucasians, for potential prevention, reduced disease progression, or severity.

15.2.2.3 Viruses

Epstein Barr Virus (EBV), a herpesvirus transmitted through saliva and other secretions, is implicated in MS by replicating in B cells and triggering B cell proliferation (Hatton et al. 2014). EBV establishes latency in its target cells, where several proteins produced by the virus can manipulate the host cell's life cycle and potentially trigger B-cell lymphomas. Like most viruses, EBV triggers a T helper (Th)-1 cell-mediated immune response and encodes antiviral cytokines, including interferon-alpha and interferon-gamma. The virus also produces proteins that interfere with the HLA loading of viral antigens and inhibit apoptosis of immune cells, thus manipulating the host's immune response to its advantage (Hatton et al. 2014). Studies have shown that the risk of MS is much higher in individuals with a history of symptomatic EBV, specifically infectious mononucleosis, compared to those without this history (Ascherio and Munger 2007, 2010). Further, there is a demonstrated association between MS and the presence of antibodies and T cells responding to lytic phase proteins of EBV, which is indicative of a recent infection or re-activation (Tarlinton et al. 2019). In summary, EBV has been shown to be associated with MS; however, further research is required to understand the efficacy of EBV treatment in preventing or treating MS.

15.3 Epigenetic Mechanisms of Gene Regulation

DNA is packaged in the nucleus into chromatin, a dynamic protein-DNA complex consisting of DNA, histones, and non-histone proteins (Kouzarides 2007; Luger et al. 1997). The nucleosome is the basic subunit of chromatin and serves as the foundation for epigenetic modifications, altering chromatin structure through DNA modifications and nucleosome rearrangement (Jenuwein and Allis 2001). Chromatin exists in two primary states: euchromatin and heterochromatin. Euchromatin, characterized by its loose coiling, allows transcription factors to interact with gene regulatory regions, such as promoters and enhancers. In contrast, heterochromatin is a tightly coiled and compressed chromatin that prevents protein-DNA interactions (van den Elsen et al. 2014). The mechanisms of epigenetics are a crucial factor in the transcriptional regulation of genes, maintenance of cellular identity, cell activation, repair, and stress processes (van den Elsen et al. 2014). The role of epigenetics in gene regulation is influenced by three major mechanisms: DNA methylation, histone code modifications, and microRNA-regulated gene expression (Oksenberg 2013).

15.3.1 DNA Methylation

Methylation is a process by which a methyltransferase enzyme covalently links a methyl group to a DNA molecule or protein substrate through an energy-dependent reaction (Friso et al. 2017). In eukaryotes, DNA methylation occurs when a methyl group is transferred to the fifth carbon of a pyrimidine ring on a cytosine via the action of DNA methyltransferases [DNMTs] (Flores et al. 2013). The process primarily occurs in CpG dinucleotides, which are commonly found in regulatory and promoter regions (Weber et al. 2007). DNA methylation in promoter regions induces the formation of heterochromatin, a tightly packed form of DNA, that prevents access to transcriptional machinery (Flores et al. 2013). Thus, the addition of methyl groups in the promoter region contributes to gene silencing (Weber et al. 2007). Additionally, DNA methylation can result in structural changes to chromatin by attracting protein complexes that modify the histone scaffold holding the DNA coil (Flores et al. 2013).

Mammals primarily have three main DNMTs, DNMT1, DNMT3a, and DNMT3b, each harboring a unique function (Celarain and Tomas-Roig 2020). DNMT1 plays a role in the DNA repair system and contributes to the cell cycle, specifically in maintaining the DNA methylation pattern (Mortusewicz et al. 2005). DNMT3a and DNMT3b catalyze the de novo addition of a methyl group into a naked cytosine (Van Emburgh and Robertson 2011). Collectively, this process is complemented by the collaboration with specific transcription factors or binding of transcription factors, resulting in the methylation of all CpG sites (Lienert et al. 2011).

15.3.2 Histone Code Modifications

Histones are highly conserved proteins that act as building blocks of nucleosomes, the fundamental structural and functional unit of chromatin (He et al. 2018). The nucleosome is wrapped by DNA, consisting of two copies of four core histones (H2A, H2B, H3, and H4) and linked by histone H1 (He et al. 2018; Luger et al. 1997)]. The five classes of histone proteins have over 60 different residues that constitute the major protein component of chromatin and further allow for tight DNA packing (He et al. 2018). In addition, the flexible N-terminus, referred to as the "histone tail," is a crucial component that undergoes post-translational modifications, such as acetylation (Luger et al. 1997). These modifications allow regulatory proteins to access DNA, thereby influencing chromatin-mediated processes like chromatin condensation, gene transcription, DNA replication, and DNA damage repair (He et al. 2018; Önder et al. 2015).

Histones can undergo post-translational modification through chemical modifications (He et al. 2018). The histone tail, flexible N-terminus domain, contains multiple lysine residues that are susceptible to acetylation, methylation, and phosphorylation (He et al. 2018; Tessarz and Kouzarides 2014). Histone acetylation occurs via two different mechanisms: the direct and effector-mediated readout mechanisms, each contributing to the dynamic regulation of chromatin structure and gene expression. The direct mechanism of histone acetylation neutralizes the positive charge on lysine residues, destabilizing the DNA-histone interaction (Tessarz and Kouzarides 2014). This, in turn, permits an open, loosely packed chromatin structure that allows access to specific transcription factors and machinery.

15.3.3 MicroRNAs

MicroRNAs (miRNA) are conserved single-stranded non-coding RNAs that influence cell processes such as differentiation, development, cell cycle regulation, apoptosis, stress response, inflammation, neurogenesis, and gliogenesis (Chitnis et al. 2020; Wang 2021; Zheng et al. 2010). The transcription of genes encoding miRNAs is regulated by complex processes, with one factor being the epigenetic mechanism of DNA methylation (Kulis and Esteller 2010). In contrast, a distinct set of miRNAs control the expression of important epigenetic regulators, such as DNMTs, via posttranscriptional modifications that target mRNA degradation or translational inhibition of mRNAs encoding DNMTs (Sato et al. 2011). The functional mechanism of miRNAs involves binding to the 3′-UTR of target mRNAs by base pairing with the seed sequence in the 5′-miRNAs. This interaction either inhibits translation and elongation by degrading target mRNAs or reduces target mRNA levels by inducing destabilization (Guo et al. 2010; Li and Yao 2012).

The hallmark of MS is the initial immune-mediated inflammatory response leading to myelin injury and axonal damage at the onset of the disease (Chastain et al. 2011). Oligodendrocytes, essential neuronal cells in the central nervous system, play a crucial role in producing and maintaining the myelin sheath that surrounds axons for optimal action potential delivery (Li and Yao 2012). To initiate and maintain myelinating function, oligodendrocytes must originate from oligodendrocyte precursor cells, a process driven by a cascade of transcriptional and posttranscriptional events mediated by miRNAs, DNA-binding proteins, and chromatin regulators (Liu and Casaccia 2010; Nave 2010; Swiss et al. 2011). Histone acetylation, specifically histone H3, has been associated with increased levels of transcriptional inhibitors affecting oligodendrocyte differentiation (Li et al. 2009; Shen et al. 2008). Studies have shown that miRNAs appear to influence both oligodendrocyte differentiation and the development of the immune system. Specifically, miR-223, a miRNA known to modulate the Nuclear factor kappa B (NF-κB) pathway, has been shown to be upregulated in the blood and in regulatory T (Tregs) cells of MS patients compared to healthy controls (Keller et al. 2009; Li et al. 2010). Further, posttranslational modifications in MS result in hypomethylation of the peptidyl arginine deaminase 2 (PAD2) gene promoter in white matter, leading to increased PAD2 expression (Mastronardi et al. 2007; Moscarello et al. 2007). PAD2 is an enzyme that catalyzes the citrullination of myelin basic protein (MBP), leading to increased levels of citrullinated MBP, which, in turn, contributes to the loss of myelin stability in the brains of individuals with MS (Beato and Sharma 2020; D'Souza et al. 2005).

15.4 Conclusion

In conclusion, as multiple sclerosis is a disease of multifactorial origin, treating and managing the disease is a challenge. Extensive research has been dedicated to exploring the influences of MS, including genetic factors, environmental elements, and epigenetic modifications. Epigenetic modifications, in particular, have gained attention for their implications in MS development. Further, therapeutic research on MS should prioritize the three mechanisms of epigenetic modification, as current research identifies multiple factors within each mechanism that influence disease progression or recurrence. Targeting methylation and histone acetylation that occurs during MS may offer a potential therapeutic approach for countering the effects of the disease.

15.5 Summary

This chapter delves into the multifaceted nature of MS as a disease influenced by genetic, environmental, and epigenetic factors. It emphasizes the role of epigenetic modifications, including DNA methylation, histone code modifications, and microRNA-regulated gene expression, in altering gene expression without changing the DNA sequence. DNA methylation involves adding methyl groups to cytosine residues, leading to gene silencing and chromatin structure changes. Histone modifications, such as acetylation and methylation, affect chromatin accessibility and gene expression. MicroRNAs regulate gene expression post-transcriptionally by binding to target mRNAs. The dysregulation of these epigenetic mechanisms plays a crucial role in MS pathogenesis, highlighting the need for further research into these processes. Understanding these mechanisms can offer insights into potential therapeutic strategies for managing MS by targeting specific epigenetic modifications.

References

Adamczyk-Sowa M, Gębka-Kępińska B, Kępiński M (2020) Multiple sclerosis—risk factors. Wiad Lek 73(12):2677–2682
Ascherio A, Munger KL (2007) Environmental risk factors for multiple sclerosis. Part I: the role of infection. Ann Neurol 61(4):288–299
Ascherio A, Munger KL (2010) Epstein–Barr virus infection and multiple sclerosis: a review. J Neuroimmune Pharmacol 5:271–277
Beato M, Sharma P (2020) Peptidyl arginine deiminase 2 (PADI2)-mediated arginine citrullination modulates transcription in cancer. Int J Mol Sci 21(4):1351
Berger SL, Kouzarides T, Shiekhattar R, Shilatifard A (2009) An operational definition of epigenetics. Genes Dev 23(7):781–783
Borch-Johnsen K, Olsen JH, Sørensen TI (1994) Genes and family environment in familial clustering of cancer. Theor Med 15:377–386
Brola W, Sobolewski P, Szczuchniak W, Góral A, Fudala M, Przybylski W, Opara J (2016) Association of seasonal serum 25-hydroxyvitamin D levels with disability and relapses in relapsing-remitting multiple sclerosis. Eur J Clin Nutr 70(9):995–999
Browne P, Chandraratna D, Angood C, Tremlett H, Baker C, Taylor BV, Thompson AJ (2014) Atlas of multiple sclerosis 2013: a growing global problem with widespread inequity. Neurology 83(11):1022–1024

Celarain N, Tomas-Roig J (2020) Aberrant DNA methylation profile exacerbates inflammation and neurodegeneration in multiple sclerosis patients. J Neuroinflammation 17(1):1–17

Chastain EM, d'Anne SD, Rodgers JM, Miller SD (2011) The role of antigen presenting cells in multiple sclerosis. Biochim Biophys Acta 1812(2):265–274

Chitnis T, Prat A, Mycko MP, Baranzini SE (2020) microRNA and exosome profiling in multiple sclerosis. Mult Scler J 26(5):599–604

Compston A, Winedl H, Kieseier B (2008) Coles. Multiple sclerosis. Lancet 372:1502–1517

D'Souza CA, Wood DD, She Y-M, Moscarello MA (2005) Autocatalytic cleavage of myelin basic protein: an alternative to molecular mimicry. Biochemistry 44(38):12905–12913

Di Majo D, Cacciabaudo F, Accardi G, Gambino G, Giglia G, Ferraro G, Candore G, Sardo P (2022) Ketogenic and modified mediterranean diet as a tool to counteract neuroinflammation in multiple sclerosis: nutritional suggestions. Nutrients 14(12):2384

Donnan PT, Parratt JD, Wilson SV, Forbes RB, O'Riordan JI, Swingler RJ (2005) Multiple sclerosis in Tayside, Scotland: detection of clusters using a spatial scan statistic. Mult Scler J 11(4):403–408

Doshi A, Chataway J (2016) Multiple sclerosis, a treatable disease. Clin Med 16(6):s53–s59. https://doi.org/10.7861/clinmedicine.16-6-s53

Dyment D, Yee I, Ebers G, Sadovnick A (2006) Multiple sclerosis in stepsiblings: recurrence risk and ascertainment. J Neurol Neurosurg Psychiatry 77(2):258

Flores KB, Wolschin F, Amdam GV (2013) The role of methylation of DNA in environmental adaptation. Oxford University Press, Oxford

Friso S, Udali S, De Santis D, Choi S-W (2017) One-carbon metabolism and epigenetics. Mol Asp Med 54:28–36

Gianfrancesco MA, Stridh P, Rhead B, Shao X, Xu E, Graves JS, Chitnis T, Waldman A, Lotze T, Schreiner T (2017) Evidence for a causal relationship between low vitamin D, high BMI, and pediatric-onset MS. Neurology 88(17):1623–1629

Guo H, Ingolia NT, Weissman JS, Bartel DP (2010) Mammalian microRNAs predominantly act to decrease target mRNA levels. Nature 466(7308):835–840

Hatton OL, Harris-Arnold A, Schaffert S, Krams SM, Martinez OM (2014) The interplay between Epstein–Barr virus and B lymphocytes: implications for infection, immunity, and disease. Immunol Res 58:268–276

Häusser-Kinzel S, Weber MS (2019) The role of B cells and antibodies in multiple sclerosis, neuromyelitis optica, and related disorders. Front Immunol 10:201

He H, Hu Z, Xiao H, Zhou F, Yang B (2018) The tale of histone modifications and its role in multiple sclerosis. Hum Genomics 12(1):1–12

Hemminki K, Li X, Sundquist J, Hillert J, Sundquist K (2009) Risk for multiple sclerosis in relatives and spouses of patients diagnosed with autoimmune and related conditions. Neurogenetics 10:5–11

Jenuwein T, Allis CD (2001) Translating the histone code. Science 293(5532):1074–1080

Keller A, Leidinger P, Lange J, Borries A, Schroers H, Scheffler M, Lenhof H-P, Ruprecht K, Meese E (2009) Multiple sclerosis: microRNA expression profiles accurately differentiate patients with relapsing-remitting disease from healthy controls. PLoS ONE 4(10):e7440

Kouzarides T (2007) Chromatin modifications and their function. Cell 128(4):693–705

Kulis M, Esteller M (2010) DNA methylation and cancer. In: Herceg Z, Ushijima T (eds) Advances in genetics, vol 70. Academic, New York, pp 27–56. https://doi.org/10.1016/B978-0-12-380866-0.60002-2

Li J-S, Yao Z-X (2012) MicroRNAs: novel regulators of oligodendrocyte differentiation and potential therapeutic targets in demyelination-related diseases. Mol Neurobiol 45:200–212

Li H, He Y, Richardson WD, Casaccia P (2009) Two-tier transcriptional control of oligodendrocyte differentiation. Curr Opin Neurobiol 19(5):479–485

Li T, Morgan MJ, Choksi S, Zhang Y, Kim Y-S, Liu Z-G (2010) MicroRNAs modulate the noncanonical transcription factor NF-κB pathway by regulating expression of the kinase IKKα during macrophage differentiation. Nat Immunol 11(9):799–805

Lienert F, Wirbelauer C, Som I, Dean A, Mohn F, Schübeler D (2011) Identification of genetic elements that autonomously determine DNA methylation states. Nat Genet 43(11):1091–1097

Liu J, Casaccia P (2010) Epigenetic regulation of oligodendrocyte identity. Trends Neurosci 33(4):193–201

Lublin FD (2014) New multiple sclerosis phenotypic classification. Eur Neurol 72(1):1–5

Luger K, Mäder AW, Richmond RK, Sargent DF, Richmond TJ (1997) Crystal structure of the nucleosome core particle at 2.8 Å resolution. Nature 389(6648):251–260

Mastronardi FG, Noor A, Wood DD, Paton T, Moscarello MA (2007) Peptidyl argininedeiminase 2 CpG island in multiple sclerosis white matter is hypomethylated. J Neurosci Res 85(9):2006–2016

Mortusewicz O, Schermelleh L, Walter J, Cardoso MC, Leonhardt H (2005) Recruitment of DNA methyltransferase I to DNA repair sites. Proc Natl Acad Sci 102(25):8905–8909

Moscarello MA, Mastronardi FG, Wood DD (2007) The role of citrullinated proteins suggests a novel mechanism in the pathogenesis of multiple sclerosis. Neurochem Res 32:251–256

Nave K-A (2010) Oligodendrocytes and the "micro brake" of progenitor cell proliferation. Neuron 65(5):577–579

Nieves J, Cosman F, Herbert J, Shen V, Lindsay R (1994) High prevalence of vitamin D deficiency and reduced bone mass in multiple sclerosis. Neurology 44(9):1687–1687

Oksenberg JR (2013) Decoding multiple sclerosis: an update on genomics and future directions. Expert Rev Neurother 13(2):11–19

Önder Ö, Sidoli S, Carroll M, Garcia BA (2015) Progress in epigenetic histone modification analysis by mass spectrometry for clinical investigations. Expert Rev Proteomics 12(5):499–517

Ostkamp P, Schwab N (2022) Effects of latitude and sunlight on multiple sclerosis severity: two peas in a pod? AAN Enterprises 98:997–998

Ramagopalan S, Handel A, Giovannoni G, Siegel SR, Ebers GC, Chaplin G (2011) Relationship of UV exposure to prevalence of multiple sclerosis in England. Neurology 76(16):1410–1414

Sadovnick AD, Baird PA, Ward RH, Optiz JM, Reynolds JF (1988) Multiple sclerosis. Updated risks for relatives. Am J Med Genet 29(3):533–541

Sato F, Tsuchiya S, Meltzer SJ, Shimizu K (2011) MicroRNAs and epigenetics. FEBS J 278(10):1598–1609

Shen S, Sandoval J, Swiss VA, Li J, Dupree J, Franklin RJ, Casaccia-Bonnefil P (2008) Age-dependent epigenetic control of differentiation inhibitors is critical for remyelination efficiency. Nat Neurosci 11(9):1024–1034

Simpson S, Blizzard L, Otahal P, Van der Mei I, Taylor B (2011) Latitude is significantly associated with the prevalence of multiple sclerosis: a meta-analysis. J Neurol Neurosurg Psychiatry 82(10):1132–1141

Swiss VA, Nguyen T, Dugas J, Ibrahim A, Barres B, Androulakis IP, Casaccia P (2011) Identification of a gene regulatory network necessary for the initiation of oligodendrocyte differentiation. PLoS ONE 6(4):e18088

Tarlinton RE, Khaibullin T, Granatov E, Martynova E, Rizvanov A, Khaiboullina S (2019) The interaction between viral and environmental risk factors in the pathogenesis of multiple sclerosis. Int J Mol Sci 20(2):303

Tessarz P, Kouzarides T (2014) Histone core modifications regulating nucleosome structure and dynamics. Nat Rev Mol Cell Biol 15(11):703–708

van den Elsen PJ, van Eggermond MC, Puentes F, van der Valk P, Baker D, Amor S (2014) The epigenetics of multiple sclerosis and other related disorders. Mult Scler Relat Disord 3(2):163–175

Van Emburgh BO, Robertson KD (2011) Modulation of Dnmt3b function in vitro by interactions with Dnmt3L, Dnmt3a and Dnmt3b splice variants. Nucleic Acids Res 39(12):4984–5002

Vitkova M, Diouf I, Malpas C, Horakova D, Kubala Havrdova E, Patti F, Ozakbas S, Izquierdo G, Eichau S, Shaygannejad V (2022) Association of latitude and exposure to ultraviolet B radiation with severity of multiple sclerosis: an international registry study. Neurology 98(24):e2401–e2412

Wang H (2021) Micrornas, multiple sclerosis, and depression. Int J Mol Sci 22(15):7802

Weber M, Hellmann I, Stadler MB, Ramos L, Pääbo S, Rebhan M, Schübeler D (2007) Distribution, silencing potential and evolutionary impact of promoter DNA methylation in the human genome. Nat Genet 39(4):457–466

Willer C, Dyment D, Risch N, Sadovnick A, Ebers G (2003) Twin concordance and sibling recurrence rates in multiple sclerosis. Proc Natl Acad Sci 100(22):12877–12882

Zheng K, Li H, Zhu Y, Zhu Q, Qiu M (2010) MicroRNAs are essential for the developmental switch from neurogenesis to gliogenesis in the developing spinal cord. J Neurosci 30(24):8245–8250

Chapter 16
Therapeutic Strategies and Ongoing Research

Azhar Abdukadir, Nadia Rabeh, Sara Aljoudi, Zakia Dimassi, Khalood Mohamed Alhosani, and Hamdan Hamdan

Abstract Multiple sclerosis (MS), a multifaceted immunological disorder affecting over 2 million individuals globally, results in varied neurological symptoms and diminished quality of life. Diverse therapeutic modalities include disease-modifying therapies (DMTs), immunomodulators, stem cell transplantation, and neuroprotective strategies. DMTs, encompassing interferons and fumarates, aim to modulate immune responses, while novel approaches like Bruton Tyrosine Kinase (BTK) inhibitors show potential in targeting B cells and microglia. Stem cell transplantation, notably autologous hematopoietic stem cell (AHSC) transplantation, garners attention with clinical trials demonstrating superiority over conventional treatments. Additionally, neuroprotective strategies focusing on remyelination and oxidative stress reduction offer promising avenues for MS management. Challenges and prospects in MS research emphasize the necessity for personalized medicine, robust biomarkers, and innovative therapies targeting underlying disease mechanisms. Despite advancements, addressing the progressive stage of MS remains a formidable task, highlighting the ongoing imperative for novel therapeutic innovations. The current intricate landscape of MS treatment underscores the continued need for research to enhance outcomes for affected individuals.

Keywords Multiple sclerosis (MS) · Disease-modifying therapies (DMTs) · Immunomodulatory approaches · Stem cell transplantation

---

A. Abdukadir · N. Rabeh · S. Aljoudi · K. M. Alhosani · H. Hamdan (✉)
Department of Biological Sciences, College of Medicine and Health Sciences, Khalifa University, Abu Dhabi, United Arab Emirates
e-mail: hamdan.hamdan@ku.ac.ae

Z. Dimassi
Department of Medical Sciences, College of Medicine and Health Sciences, Khalifa University, Abu Dhabi, United Arab Emirates

© The Author(s), under exclusive license to Springer Nature Singapore Pte Ltd. 2024
H. Hamdan (ed.), *Exploring the Effects of Diet on the Development and Prognosis of Multiple Sclerosis (MS)*, Nutritional Neurosciences, https://doi.org/10.1007/978-981-97-4673-6_16

Abbreviations

AHSC	Autologous hematopoietic stem cell
AI	Artificial intelligence
ASTIMS	Autologous hematopoietic stem cell transplantation trial in MS
BBB	Blood-brain barrier
BTK	Bruton tyrosine kinase
CARs	Chimeric antigen receptors
CLL	Chronic lymphocytic leukemia
CNS	Central nervous system
DMTs	Disease-modifying therapies
IFNs	Interferons
Ig	Immunoglobulin
IPSC	Induced pluripotent stem cells
LINGO-1	Leucine-rich repeat and immunoglobulin domain-containing Nogo receptor-interacting protein 1
MBP	Myelin basic protein
MIST	Multiple sclerosis international stem cell transplant
MRI	Magnetic resonance imaging
MS	Multiple sclerosis
NK	Natural killer
PML	Progressive multifocal leukoencephalopathy
rHIgM22	Recombinant human immunoglobulin M22
RRMS	Relapsing-remitting MS
S1P	Sphingosine 1 phosphate
Th	T helper
Treg	Regulatory T
VCAM-1	Vascular cell adhesion molecule 1

Learning Objectives
- Describe the different classes of DMTs used in the treatment of MS, including their mechanisms of action, benefits, and challenges.
- Identify various immunomodulatory therapies, such as interferons, fumarates, and monoclonal antibodies, and explain how they modulate the immune response in MS patients.
- Analyze the role of autologous hematopoietic stem cell (AHSC) transplantation and induced pluripotent stem cells (IPSC) in MS treatment, including the potential benefits and associated risks.
- Understand the importance of neuroprotective strategies in MS management, focusing on remyelination and oxidative stress reduction, and discuss promising therapeutic agents.

- Assess the future directions of MS research, emphasizing the need for personalized medicine, reliable biomarkers, and innovative therapies targeting the underlying disease mechanisms.

16.1 Introduction

Multiple sclerosis (MS) is a chronic immunological disorder that manifests as a wide range of neurological symptoms, therapeutic responses, and neuropathological aspects. Demyelination and neuroinflammation are side effects of MS that are harmful to the patient. Over 2 million people worldwide are affected by MS during their most productive years, with patients experiencing a wide spectrum of symptoms that can significantly lower their quality of life (Walton et al. 2020). Understanding the many stages of MS, from asymptomatic to radiologically isolated condition, clinically isolated syndrome, relapsing-remitting MS, and other progressive forms, is critical (Walton et al. 2020). While no definitive treatment for MS has been yet determined, significant strides have been made to understand its underlying mechanisms and create therapeutic approaches.

Current therapies that aim to ameliorate the long-term outcomes of MS are classified as disease-modifying therapies (DMTs). DMTs are divided into several groups, namely immunomodulators, stem cell transplantation, and neuroprotective strategies (Piehl 2021). Before being approved for clinical use, each drug undergoes a rigorous evaluation through clinical trials. The success of these treatments is monitored through radiological or biochemical methods (Piehl 2021). The current standard for clinical monitoring is magnetic resonance imaging (MRI), which probes for demyelinating central nervous system (CNS) lesions. With the advent of artificial intelligence (AI), AI-assisted MRI monitoring and diagnosis of MS has reached a superior diagnostic sensitivity of 93.3% compared to radiological reports (Barnett et al. 2023). Developing increasingly sensitive monitoring tools is imperative as MS diagnosed at a later stage can manifest with primarily irreversible damage to the CNS (Piehl 2021). Despite advancements in MS monitoring and a growing library of DMTs, validated guidelines for treatment monitoring and algorithms are still lacking. This chapter will discuss the current DMTs approved for MS treatment and future outlooks.

16.2 Immunomodulators

There are countless novel mechanisms of action for MS, so investigating therapies that target different aspects of the immune response will help combat the disease (Cross and Naismith 2014). One mechanism targets the modulation of cytokine activity during the inflammatory state, potentially dampening the neuroinflammatory processes seen in MS, leading to symptom relief and halting disease progression (Kantarci et al. 2014). Interferons (IFNs), specifically IFN-β-1a are one of the first class of immunomodulators to be approved for treating MS patients. IFNs, through their interactions with lymphocytes and antigen-presenting cells, promote an anti-inflammatory state. Earlier formulations were unattractive due to their side-effect profiles, but more recent formulations are a pegylated form of IFN-β-1a, which had moderate success in reducing annual relapse rates. All IFN therapies are in the form of subcutaneous injections, leading to poor adherence to therapy. However, the pegylated form of IFN-β-1a is superior to other injectable IFN treatments as it requires fewer doses (Calabresi et al. 2014). A class of drugs with a similar outcome to that of IFN therapy is fumarates, which are derived from fumaric acid and possess an immunomodulatory potential. There is not a single method in which fumarates induce an anti-inflammatory state, but rather a series of effects (Piehl 2021). For example, dimethyl fumarate, a drug taken orally for MS treatment, promotes the T helper (Th)-2 cell's response to induce an anti-inflammatory state (Bomprezzi 2015).

Several oral and infusion therapies have succeeded in modulating the activity of immune system cells, which led to reductions in relapses and disease progression. One target of such therapies is the B cells, immune cells that contribute to the autoimmune response by producing autoantibodies (Hoffman et al. 2016). B cells are never present in the central nervous system (CNS) since they permeate the blood-brain barrier (BBB). Due to inflammatory reactions that deteriorate the BBB's integrity, B cells infiltrate the CNS of patients with MS, leading to the destruction of the myelin sheath and the neurons they cover (Cross and Naismith 2014). Another valuable target for MS therapy is the CD20 protein expressed on the surface of B cells, which has been implicated in the pathogenesis of MS by attacking the myelin sheath. For example, Ocrelizumab, an infusion treatment authorized for both relapsing and primary progressive MS, is a monoclonal antibody that lyses particular B cells, namely CD20+ B cells. Depleting CD20+ B cells reduces inflammation and halts the demyelination of neurons (Castro-Borrero et al. 2012). Two similar drugs, Ofatumumab and Ublituximab, also target CD20+ B cells and help reduce the autoimmune response in MS by decreasing B cells and changing their activity (Cross and Naismith 2014). Ofatumumab is administered subcutaneously every month, while Ublituximab and Ocrelizumab are given as infusions approximately twice a year (Fox et al. 2020; Piehl 2021). Although these therapies are generally effective, their differences in administration and dosing could significantly influence a patient's adherence to the treatment regimen. Alemtuzumab, a selective humanized monoclonal antibody directed against the CD52 antigen expressed on CD4+ and CD8+ T cells, B cells, monocytes, and dendritic cells, has the advantage of affecting both lymphoid and myeloid cells, ultimately inducing lysis of these cells (Guarnera et al. 2017). Despite its therapeutic potential, Alemtuzumab causes major side effects such as thyroid cancer, lymphoma, and melanoma (Cross and Naismith 2014).

Along with targeting the immune cells implicated in the pathogenesis of MS, several therapeutic strategies have indirectly targeted these cells by impeding their transport into the CNS. Fingolimod, the first approved oral therapy for MS, differs from Ocrelizumab in that it acts on both B and T cells to inhibit the lipid Sphingosine 1 phosphate (S1P) expressed on lymphocytes, which helps in T-cell trafficking (Sharma et al. 2011; Subei and Cohen 2015). It exerts its effects on the S1P receptor in lymphocytes, causing the internalization of the S1P receptor. Consequently, it prevents the migration of lymphocytes to the CNS, thereby limiting lymphocytic infiltration and inflammation (Subei and Cohen 2015). Sphingosine Kinase Inhibitors are given to further enhance the effects of Fingolimod, which block the enzymes that synthesize S1P, lowering S1P levels in the blood and modifying immune cell trafficking (Subei and Cohen 2015). Compounds sharing the same mechanism of action as Fingolimod are Ozanimod and Siponimod. Although S1P inhibitors are effective in improving the status of MS patients, they are associated with a list of adverse effects. Due to the lack of research comparing these S1P inhibitors, whether one is superior to the other is difficult to conclude (Lassiter et al. 2020).

An important target for treating MS is the prevention of the extravasation of reactive lymphocytes into the CNS. During this process, lymphocytes must adhere to endothelial cells by attaching to vascular cell adhesion molecule 1 (VCAM-1) found on the surface of blood vessels. Alpha-4 integrin potentiates this leukocyte-endothelium interaction by attaching to VCAM-1 and the fibronectin in the extracellular matrix (Kawamoto et al. 2012). Natalizumab is a drug that acts to impede the transport of lymphocytes. Blocking the alpha-4 integrin interactions between VCAM-1 and fibronectin prevents the extravasation of lymphocytes into the CNS (Brandstadter and Katz Sand 2017). A serious adverse effect associated with Natalizumab is the development of progressive multifocal leukoencephalopathy. Recently, biosimilar natalizumab, PB006, has been developed and FDA-approved for treating relapsing MS. A randomized, double-blind, single-dose clinical trial found that PB006 was as effective as natalizumab (Wessels et al. 2023). Although the study reports no patients developing progressive multifocal leukoencephalopathy (PML), PB006 is still considered a risk factor for PML development as it shares a similar side effect profile to that of natalizumab.

Other indirect approaches to lowering the abundance of T cells can help mitigate the inflammatory response seen in MS patients. One drug that exerts its effects through this mechanism is Daclizumab, an anti-CD25 antibody that binds to and enhances the activity of Natural Killer (NK) cells, leading to T cell lysis. However, due to safety concerns, it was taken off the market (Castro-Borrero et al. 2012). Lastly, cladribine, a purine antimetabolite commonly used as a chemotherapeutic agent, has been found to be successful in treating MS (Oreja-Guevara et al. 2023; Piehl 2021). It primarily acts by targeting CD4+ and CD8+ cells. Although cladribine has successfully led to a reduction in annual relapse rates, its long-term safety profile and dosing regimen remain to be confirmed.

Tolerance-inducing therapies investigate methods to induce immune tolerance to stop the autoimmune response. Galtiramer acetate is one of the oldest drugs used as a tolerance-inducing therapy in MS patients. This amino acid polymer composed of L-tyrosine, L-alanine, L-lysine, and L-glutamic acid allows it to mimic myelin basic protein (MBP), which is commonly attacked by antibodies in patients with MS. By acting as a decoy for the antibodies against MBP, there is a resultant decrease in the inflammatory response in MS patients (Piehl 2021; Weinstock-Guttman et al. 2017). Another potential target is dendritic cells, which are essential to immune responses. Tolerogenic dendritic cells are designed to increase immunological tolerance and decrease inflammation.

A newer class of drugs are those that enhance the function of regulatory T (Treg) cells to dampen the autoimmune response and reduce inflammation. One method of Treg treatment includes isolating and expanding a patient's own Treg outside the body before reinfusing it back into the patient. This procedure boosts the amount of functional Tregs, which can aid in suppressing autoimmune reactions (Verreycken et al. 2022). In a phase 1 clinical trial investigating the effectiveness of Tregs in participants with relapsing-remitting MS (RRMS), it was found that intrathecal administration halted disease progression, while relapses and progressive deterioration in status were experienced upon intravenous administration (Chwojnicki et al. 2021). Another method is to develop Treg with Chimeric Antigen Receptors (CARs), which are Treg that have been genetically modified to have chimeric antigen receptors, allowing them to target specific autoimmune triggers or antigens. There remains a need to explore the effectiveness of Tregs and CAR Tregs in the MS setting; only then can proper dosing regimens and safety profiles be elucidated.

Generally, the selection of a therapeutic regimen is determined by several variables, including the kind of MS, disease activity, unique patient characteristics, and treatment objectives. Based on each patient's unique needs and preferences, healthcare professionals collaborate closely with patients to choose the best course of therapy. Additionally, the MS treatment landscape continues to evolve, with ongoing research and development efforts to provide more effective and convenient therapies for individuals living with MS.

More recently, the use of a new class of immunomodulators for MS treatment has recently gained interest. Bruton Tyrosine Kinase (BTK) inhibitors work on the intracellular BTK receptor, which is involved in modulating the growth and activity of microglia and B cells (Krämer et al. 2023). Therefore, inhibition of microglia and B cell activity, two cells involved in MS pathogenesis, can improve the presentation of MS. What is incredibly unique is BTK inhibitors' ability to cross the BBB, allowing them to target cells within and beyond the CNS. A library of BTK inhibitors has been developed but primarily intended to treat B-cell malignancies such as chronic lymphocytic leukemia (CLL) (Krämer et al. 2023). In addition to treating malignancies, BTK inhibitors have also shown promise in treating autoimmune diseases, such as MS.

Preclinical research has investigated the efficacy of BTK inhibitors in treating MS (Krämer et al. 2023). The most commonly investigated drugs in preclinical studies include evobrutinib and tolebrutinib. These compounds act by reducing B cell growth and activity more so than T cells, as well as by preventing the infiltration and proliferation of lymphocytes and microglia (Krämer et al. 2023; Mangla et al. 2004; Pellerin et al. 2021; Torke et al. 2020). In addition to BTK inhibitors' immunosuppressive role, they have also been found to prevent demyelination and promote remyelination. A novel BTK inhibitor that is still being experimentally tested is BTKi-1, which was found to promote remyelination in vivo and ex vivo (Martin et al. 2020). Another experimental compound with properties similar to tolebrutinib, PRN2675, was found to inhibit demyelination and loss of oligodendrocytes. PRN2675 was also able to prevent the microglia migration to demyelinated sites, thereby halting the progression of the disease (Barr et al. 2020). The majority of the data is preliminary and requires peer review, meaning that the results of these studies should be interpreted with caution.

16.3 Stem Cell Transplantation

Induced Pluripotent Stem Cells (IPSC) is a new regenerative medicine groundbreaking discovery that reprograms somatic cells to express and differentiate into any cell type in the body, including neurons (Gerdoni et al. 2007). The study by Xie et al. used IPSC for their self-renewal and differentiation ability. The IPSCs were differentiated into oligodendrocytes, which primarily act to myelinate neurons, and were injected into mice models of MS. The data showed the demyelinated neurons started to regenerate their myelin and repair damaged neurons (Xie et al. 2016). In addition to the therapeutic potential of IPSC, the advantage of the application of these stem cells is their cost-effectiveness (Thanaskody et al. 2022). IPSCs derived from MS patients offer a unique platform for studying disease mechanisms. Researchers can differentiate IPSC into relevant cell types, such as oligodendrocytes, astrocytes, and neurons, to model the disease in vitro. This enables the study of MS-specific cellular and molecular abnormalities. Moreover, IPSC can be used to model MS pathology, allowing researchers to elucidate the underlying mechanisms of MS. IPSC-based models also provide a high-throughput platform for testing potential MS therapies. Researchers can use patient-derived IPSC to screen drugs for their effectiveness in rescuing disease-associated phenotypes. This approach may accelerate the development of novel therapeutics.

Numerous challenges need to be considered when using IPSC. While IPSCs offer significant promise in MS research and therapy development, the efficient and reliable differentiation of IPSCs into specific neural cell types remains a technical challenge. There remains a need to develop standardized protocols for IPSC differentiation, which is necessary to ensure consistency and reproducibility of data in the literature (Medvedev et al. 2010). Another primary concern is that IPSC overexpression could lead to the formation of teratomas. MS is a multifaceted disease, and recapitulating the full complexity of the disease in vitro is a difficult task. From an ethical point of view, using IPSC raises concerns related to patient consent, genetic modification, and intellectual property. Overall, IPSC technology holds significant promise in advancing our understanding of MS pathology, modeling the disease, and developing personalized therapies. As IPSC-based techniques are further developed, researchers may gain a new understanding of MS and create ground-breaking therapies that enhance the lives of those who suffer from this crippling condition.

Collaboration among scientists, clinicians, and regulatory bodies is essential to address the challenges associated with IPSC-based research and translation into clinical applications.

Mounting evidence for autologous hematopoietic stem cell (AHSC) transplantation in MS patients has shown the potential to halt disease progression. Compared to DMTs, AHSCs were found to be highly cost-effective in the long term, especially in aggressive cases of MS (Burt et al. 2020; Mariottini et al. 2024). The AHSC transplant process begins with a "wash-out" phase, where all immune-modulating drugs and DMTs are discontinued. The significance of the "wash-out" phase is to reduce the risk of potential inhibitory effects on the AHSC transplant. The "wash-out" phase is also said to be routinely used in practice when switching patients from one DT to another. These measures are taken to reduce the risk of PML. Once the "wash-out" phase is complete, HSCs from the peripheral blood are harvested through leukapheresis, a process by which leukocytes are isolated from other components of the blood. During and before the start of this process, approximately 2 g/m^2 of cyclophosphamide is administered as an immunosuppressant and method to immobilize stem cells from the bone marrow (Mohammadi et al. 2021). Due to limited data, long-term adverse effects of MS patients remain to be elucidated. However, some studies cite impaired fertility and reproductive system function, development of autoimmunity, and endocrine disorders as long-term adverse effects.

Several clinical trials are underway to investigate AHSC transplantation's efficacy in MS patients. A few clinical trials with published results highlight the advantages of AHSC transplantation. In a phase II randomized control trial termed ASTIMS (Autologous Hematopoietic Stem Cell Transplantation Trial in MS), the efficacy of AHSC transplants was compared to patients taking mitoxantrone, an antineoplastic drug. The trial found the AHSC transplants superior to the regular mitoxantrone therapy. However, the results should be interpreted with caution due to the very small sample size. The MIST (Multiple Sclerosis International Stem Cell Transplant) trial compares AHSC transplantation to DMTs. Unlike the ASTIMS trial, the MIST trial was conducted on a larger scale. The trial results found AHSC transplants superior to DMTs when treating patients with MS (Burt et al. 2019). Evidently, AHSC transplantation carries significant therapeutic potential. Despite these positive findings, there remains a need for additional clinical trials to explore the effect of various AHSC transplantation methods on MS activity, as the optimal method can pave the way for safe and effective transplantation. Increasingly robust trials are needed to compare AHSC transplantation with existing DMTs. Results from these trials would further confirm the advantage of AHSC transplant in MS treatment. It is also worth exploring the effect of AHSC transplantation in patients with MS at different stages of the disease. The positive clinical trial outcomes underscore the necessity for collaborative research to address remaining knowledge gaps.

16.4 Neuroprotective Strategies

Neuroprotective research seeks to preserve and repair damaged neurons. A way to restore neurons to their original, healthy state is by inducing remyelination. Remyelination is a natural healing process involving the regeneration of myelin sheaths around injured nerve axons. This procedure can return nerve transmission to normal and preserve nerve axons from future injury (Greenberg et al. 2022). The amount of remyelination in MS varies between individuals. It can deteriorate with time, particularly in progressive types of MS. Developing drugs and strategies to promote remyelination and restore neural function is challenging for several reasons. First, it is difficult to differentiate between intact white matter and white matter formed due to remyelination (Neumann et al. 2020). Second, since established protocols for the dating of MS lesions are lacking, it becomes difficult to establish when and how remyelination begins (Neumann et al. 2020). Despite these difficulties, therapeutic agents for the induction of remyelination have shown promising potential.

A target of interest for remyelination is the leucine-rich repeat and immunoglobulin domain-containing Nogo receptor-interacting protein 1 (LINGO-1) on the surface of oligodendrocytes. LINGO-1 is known to inhibit oligodendrocyte differentiation and myelination. Opicinumab is known to inhibit LINGO-1 and has been evaluated under several clinical trials. However, due to a lack of significant improvement in the disability status of MS patients, the development of ipilimumab was terminated (Oh and Bar-Or 2022). In a mice model where LINGO-1 was inhibited, the oligodendrocyte could differentiate, promote remyelination, and demonstrate axonal regeneration (Cross and Naismith 2014). However, a more promising therapeutic agent is the recombinant human immunoglobulin M22 (rHIgM22), a monoclonal immunoglobulin (Ig) M antibody that increases oligodendrocyte differentiation and proliferation and promotes axon remyelination (Greenberg et al. 2022). rHIgM22's promising therapeutic potential and minimal side effects warrant further research to elucidate optimal dosing regimens and mechanisms of action.

Neuroprotective research also focuses on reducing oxidative stress and preventing apoptosis of neurons and oligodendrocytes. Oxidative stress is a significant factor in cellular degradation and neurodegenerative disorders. Therefore, lowering oxidative stress preserves axonal integrity, as well as aid in tissue regeneration and neuroprotection. An experimental agent for the treatment of MS that affects the production of free radicals is clomipramine. In mice models of MS clomipramine could reduce MS severity; these results warrant further investigation in clinical trials (Oh and Bar-Or 2022).

Several obstacles must be addressed to induce successful remyelination in MS. These difficulties include inhibitory factors that impede remyelination, the patient's age, the level of axonal damage, and the disease's chronicity. Research is underway to improve our understanding of remyelination and the neuroprotective agents that can potentiate this process. Furthermore, existing neuroprotective strategies offer therapeutic promise, warranting further investigation in clinical trials to confirm their safety and efficacy.

16.5 Future Outlook

The future of MS treatment relies on advancements in DMTs and their immunomodulatory therapies. The evolving landscape of DMTs in MS is characterized by a growing array of treatment options, each designed to intervene at different points in the disease process. Traditional injectable therapies have been augmented by the emergence of oral medications, such as fingolimod, and monoclonal antibodies, like ocrelizumab. Looking forward, the development of DMTs is expected to be increasingly informed by a deeper understanding of the molecular mechanisms underpinning MS. Specific molecular targets within the immune system and the CNS are being explored, and therapeutic strategies are being designed to modulate these targets with precision. Moreover, the advent of personalized medicine in MS treatment involves tailoring DMTs based on a patient's genetic and molecular profile. This could involve identifying genetic markers associated with treatment response or susceptibility to particular side effects, allowing for a more individualized and effective therapeutic approach.

In the new age of precision medicine, there is a call for the development of reliable biomarkers. These biomarkers can indicate disease activity, predict treatment response, and guide therapeutic decision-making. Identifying specific biomarkers indicative of disease activity and treatment response could enable healthcare providers to customize treatment plans for individual patients. This personalized medicine approach ensures patients receive the most effective therapy based on their unique immune profile.

In summary, the future outlook for immunomodulatory therapies and DMTs in MS treatment involves a meticulous dissection of the immune system's intricacies and a move toward personalized medicine. The aim is to strike a balance between efficacy and safety, providing individuals with MS treatments that are highly effective in managing the disease and tailored to their unique biological characteristics. As research continues to unravel the complexities of MS, the therapeutic landscape is poised for transformative changes, offering new hope and improved outcomes for those affected by this challenging condition.

16.6 Conclusion

Even though there are no definitive guidelines for the treatment of MS, a lot of work has been done in understanding its underlying mechanisms and creating therapeutic approaches. In terms of therapies, DMTs have shown promise in lowering relapse rates and forming new lesions, but their capacity to treat the progressive stage is yet unknown. IPSC technology is promising in advancing our understanding of MS pathology, modeling the disease, and developing personalized therapies. However, the primary concern of IPSC overexpression leading to teratoma formation must be addressed. The most promising avenue is developing novel tissue-repair therapies, such as remyelination-promoting antibodies, which will best combat MS. The progressive stage of MS remains a difficult component for both patients and clinicians, highlighting the need for a paradigm shift in preclinical multiple sclerosis research to address the root causes of impairment in this disorder and hopefully one of these therapies will be the final piece that solves the problem.

16.7 Summary

MS is a complex immunological disorder affecting over 2 million individuals worldwide, leading to varied neurological symptoms and reduced quality of life. This chapter explores diverse therapeutic modalities for MS, including disease-modifying therapies (DMTs), immunomodulators, stem cell transplantation, and neuroprotective strategies. DMTs like interferons and fumarates aim to modulate immune responses, while novel approaches, such as Bruton Tyrosine Kinase (BTK) inhibitors, show promise in targeting B cells and microglia. Stem cell transplantation, particularly autologous hematopoietic stem cell (AHSC) transplantation, has demonstrated superiority over conventional treatments in clinical trials. Additionally, neuroprotective strategies focusing on remyelination and reducing oxidative stress offer promising avenues for managing MS. Despite advancements, addressing the progressive stage of MS remains challenging, highlighting the need for continued research and innovation in therapeutic approaches. The chapter emphasizes the importance of personalized medicine, robust biomarkers, and ongoing research to enhance outcomes for individuals affected by MS.

References

Barnett M, Wang D, Beadnall H, Bischof A, Brunacci D, Butzkueven H, Brown JWL, Cabezas M, Das T, Dugal T, Guilfoyle D, Klistorner A, Krieger S, Kyle K, Ly L, Masters L, Shieh A, Tang Z, van der Walt A, Wang C (2023) A real-world clinical validation for AI-based MRI monitoring in multiple sclerosis. NPJ Dig Med 6(1):196. https://doi.org/10.1038/s41746-023-00940-6

Barr H, Given K, Mcclain C, Gruber R, Ofengeim D, Macklin W, Hughes E (2020, December). BTK signaling regulates real-time microglial dynamics and prevents demyelination in a novel in vivo model of antibodymediated cortical demyelination. In MULTIPLE SCLEROSIS JOURNAL (Vol. 26, No. 3_ SUPPL, pp. 39-39). 1 OLIVERS YARD, 55 CITY ROAD, LONDON EC1Y 1SP, ENGLAND: SAGE PUBLICATIONS LTD

Bomprezzi R (2015) Dimethyl fumarate in the treatment of relapsing–remitting multiple sclerosis: an overview. Ther Adv Neurol Disord 8(1):20–30. https://doi.org/10.1177/1756285614564152

Brandstadter R, Katz Sand I (2017) The use of natalizumab for multiple sclerosis. Neuropsychiatr Dis Treat 13:1691–1702

Burt RK, Balabanov R, Burman J, Sharrack B, Snowden JA, Oliveira MC, Fagius J, Rose J, Nelson F, Barreira AA, Carlson K, Han X, Moraes D, Morgan A, Quigley K, Yaung K, Buckley R, Alldredge C, Clendenan A, Helenowski IB (2019) Effect of nonmyeloablative hematopoietic stem cell transplantation vs continued disease-modifying therapy on disease progression in patients with relapsing-remitting multiple sclerosis: a randomized clinical trial. JAMA 321(2):165–174. https://doi.org/10.1001/jama.2018.18743

Burt RK, Tappenden P, Han X, Quigley K, Arnautovic I, Sharrack B, Snowden JA, Hartung D (2020) Health economics and patient outcomes of hematopoietic stem cell transplantation versus disease-modifying therapies for relapsing remitting multiple sclerosis in the United States of America. Mult Scler Relat Disord 45:102404. https://doi.org/10.1016/j.msard.2020.102404

Calabresi PA, Kieseier BC, Arnold DL, Balcer LJ, Boyko A, Pelletier J, Liu S, Zhu Y, Seddighzadeh A, Hung S, Deykin A (2014) Pegylated interferon beta-1a for relapsing-remitting multiple sclerosis (ADVANCE): a randomised, phase 3, double-blind study. Lancet Neurol 13(7):657–665. https://doi.org/10.1016/S1474-4422(14)70068-7

Castro-Borrero W, Graves D, Frohman TC, Flores AB, Hardeman P, Logan D, Orchard M, Greenberg B, Frohman EM (2012) Current and emerging therapies in multiple sclerosis: a systematic review. Ther Adv Neurol Disord 5(4):205–220. https://doi.org/10.1177/1756285612450936

Chwojnicki K, Iwaszkiewicz-Grześ D, Jankowska A, Zieliński M, Łowiec P, Gliwiński M, Grzywińska M, Kowalczyk K, Konarzewska A, Glasner P, Sakowska J, Kulczycka J, Jaźwińska-Curyłło A, Kubach M, Karaszewski B, Nyka W, Szurowska E, Trzonkowski P (2021) Administration of CD4+CD25 high CD127−FoxP3+ regulatory T cells for relapsing-remitting multiple sclerosis: a phase 1 study. BioDrugs 35(1):47–60. https://doi.org/10.1007/s40259-020-00462-7

Cross AH, Naismith RT (2014) Established and novel disease-modifying treatments in multiple sclerosis. J Intern Med 275(4):350–363. https://doi.org/10.1111/joim.12203

Fox E, Lovett-Racke AE, Gormley M, Liu Y, Petracca M, Cocozza S, Shubin R, Wray S, Weiss MS, Bosco JA, Power SA, Mok K, Inglese M (2020) A phase 2 multicenter study of ublituximab, a novel glycoengineered anti-CD20 monoclonal antibody, in patients with relapsing forms of multiple sclerosis. Mult Scler J 27(3):420–429. https://doi.org/10.1177/1352458520918375

Gerdoni E, Gallo B, Casazza S, Musio S, Bonanni I, Pedemonte E, Mantegazza R, Frassoni F, Mancardi G, Pedotti R (2007) Mesenchymal stem cells effectively modulate pathogenic immune response in experimental autoimmune encephalomyelitis. Ann Neurol 61(3):219–227

Greenberg BM, Bowen JD, Alvarez E, Rodriguez M, Caggiano AO, Warrington AE, Zhao P, Eisen A (2022) A double-blind, placebo-controlled, single-ascending-dose intravenous infusion study of rHIgM22 in subjects with multiple sclerosis immediately following a relapse. Mult Scler J 8(2):20552173221091475

Guarnera C, Bramanti A, Mazzon E (2017) Alemtuzumab: a review of efficacy and risks in the treatment of relapsing remitting multiple sclerosis. Ther Clin Risk Manag 13:871–879. https://doi.org/10.2147/TCRM.S134398

Hoffman W, Lakkis FG, Chalasani G (2016) B cells, antibodies, and more. Clin J Am Soc Nephrol 11(1):137–154. https://doi.org/10.2215/CJN.09430915

Kantarci OH, Pirko I, Rodriguez M (2014) Novel immunomodulatory approaches for the management of multiple sclerosis. Clin Pharmacol Therap 95(1):32–44. https://doi.org/10.1038/clpt.2013.196

Kawamoto E, Nakahashi S, Okamoto T, Imai H, Shimaoka M (2012) Anti-integrin therapy for multiple sclerosis. Autoimmune Dis 2012:357101. https://doi.org/10.1155/2012/357101

Krämer J, Bar-Or A, Turner TJ, Wiendl H (2023) Bruton tyrosine kinase inhibitors for multiple sclerosis. Nat Rev Neurol 19(5):289–304. https://doi.org/10.1038/s41582-023-00800-7

Lassiter G, Melancon C, Rooney T, Murat A-M, Kaye JS, Kaye AM, Kaye RJ, Cornett EM, Kaye AD, Shah RJ, Viswanath O, Urits I (2020) Ozanimod to treat relapsing forms of multiple sclerosis: a comprehensive review of disease, drug efficacy and side effects. Neurol Int 12(3):89–108

Mangla A, Khare A, Vineeth V, Panday NN, Mukhopadhyay A, Ravindran B, Bal V, George A, Rath S (2004) Pleiotropic consequences of Bruton tyrosine kinase deficiency in myeloid lineages lead to poor inflammatory responses. Blood 104(4):1191–1197. https://doi.org/10.1182/blood-2004-01-0207

Mariottini A, Nozzoli C, Carli I, Landi F, Gigli V, Repice AM, Ipponi A, Cecchi M, Boncompagni R, Saccardi R, Massacesi L (2024) Cost and effectiveness of autologous haematopoietic stem cell transplantation and high-efficacy disease-modifying therapies in relapsing–remitting multiple sclerosis. Neurol Sci. https://doi.org/10.1007/s10072-024-07308-y

Martin E, Aigrot M-S, Grenningloh R, Stankoff B, Lubetzki C, Boschert U, Zalc B (2020) Bruton's tyrosine kinase inhibition promotes myelin repair. Brain Plast 5:123–133. https://doi.org/10.3233/BPL-200100

Medvedev S, Shevchenko A, Zakian S (2010) Induced pluripotent stem cells: problems and advantages when applying them in regenerative medicine. Acta Nat 2(5):18–27

Mohammadi R, Aryan A, Omrani MD, Ghaderian SMH, Fazeli Z (2021) Autologous hematopoietic stem cell transplantation (AHSCT): an evolving treatment avenue in multiple sclerosis. Biol Targets Ther 15:53–59

Neumann B, Foerster S, Zhao C, Bodini B, Reich DS, Bergles DE, Káradóttir RT, Lubetzki C, Lairson LL, Zalc B, Stankoff B, Franklin RJM (2020) Problems and pitfalls of identifying remyelination in multiple sclerosis. Cell Stem Cell 26(5):617–619. https://doi.org/10.1016/j.stem.2020.03.017

Oh J, Bar-Or A (2022) Emerging therapies to target CNS pathophysiology in multiple sclerosis. Nat Rev Neurol 18(8):466–475. https://doi.org/10.1038/s41582-022-00675-0

Oreja-Guevara C, Brownlee W, Celius EG, Centonze D, Giovannoni G, Hodgkinson S, Kleinschnitz C, Havrdova EK, Magyari M, Selchen D, Vermersch P, Wiendl H, Van Wijmeersch B, Salloukh H, Yamout B (2023) Expert opinion on the long-term use of cladribine tablets for multiple sclerosis: systematic literature review of real-world evidence. Mult Scler Relat Disord 69:104459. https://doi.org/10.1016/j.msard.2022.104459

Pellerin K, Rubino SJ, Burns JC, Smith BA, McCarl C-A, Zhu J, Jandreski L, Cullen P, Carlile TM, Li A, Rebollar JV, Sybulski J, Reynolds TL, Zhang B, Basile R, Tang H, Harp CP, Pellerin A, Silbereis J, Mingueneau M (2021) MOG autoantibodies trigger a tightly-controlled FcR and BTK-driven microglia proliferative response. Brain 144(8):2361–2374. https://doi.org/10.1093/brain/awab231

Piehl F (2021) Current and emerging disease-modulatory therapies and treatment targets for multiple sclerosis. J Intern Med 289(6):771–791. https://doi.org/10.1111/joim.13215

Sharma S, Mathur AG, Pradhan S, Singh DB, Gupta S (2011) Fingolimod (FTY720): first approved oral therapy for multiple sclerosis. J Pharmacol Pharmacother 2(1):49–51. https://doi.org/10.4103/0976-500x.77118

Subei AM, Cohen JA (2015) Sphingosine 1-phosphate receptor modulators in multiple sclerosis. CNS Drugs 29(7):565–575. https://doi.org/10.1007/s40263-015-0261-z

Thanaskody K, Jusop AS, Tye GJ, Wan Kamarul Zaman WS, Dass SA, Nordin F (2022) MSCs vs. iPSCs: potential in therapeutic applications. Front Cell Dev Biol 10:1005926. https://doi.org/10.3389/fcell.2022.1005926

Torke S, Pretzsch R, Häusler D, Haselmayer P, Grenningloh R, Boschert U, Brück W, Weber MS (2020) Inhibition of Bruton's tyrosine kinase interferes with pathogenic B-cell development in inflammatory CNS demyelinating disease. Acta Neuropathol 140(4):535–548. https://doi.org/10.1007/s00401-020-02204-z

Verreycken J, Baeten P, Broux B (2022) Regulatory T cell therapy for multiple sclerosis: breaching (blood-brain) barriers. Hum Vaccin Immunother 18(7):2153534. https://doi.org/10.1080/21645515.2022.2153534

Walton C, King R, Rechtman L, Kaye W, Leray E, Marrie RA, Robertson N, La Rocca N, Uitdehaag B, van der Mei I, Wallin M, Helme A, Angood Napier C, Rijke N, Baneke P (2020) Rising prevalence of multiple sclerosis worldwide: insights from the Atlas of MS, third edition. Mult Scler J 26(14):1816–1821. https://doi.org/10.1177/1352458520970841

Weinstock-Guttman B, Nair KV, Glajch JL, Ganguly TC, Kantor D (2017) Two decades of glatiramer acetate: from initial discovery to the current development of generics. J Neurol Sci 376:255–259. https://doi.org/10.1016/j.jns.2017.03.030

Wessels H, von Richter O, Velinova M, Höfler J, Chamberlain P, Kromminga A, Lehnick D, Roth K (2023) Pharmacokinetic and pharmacodynamic similarity of biosimilar natalizumab (PB006) to its reference medicine: a randomized controlled trial. Expert Opin Biol Ther 23(12):1287–1297. https://doi.org/10.1080/14712598.2023.2290530

Xie C, Liu Y-Q, Guan Y-T, Zhang G-X (2016) Induced stem cells as a novel multiple sclerosis therapy. Curr Stem Cell Res Ther 11(4):313–320

Printed in the United States
by Baker & Taylor Publisher Services